Introduction to Optical Components

Introduction to Optical Components

Roshan L. Aggarwal

Kambiz Alavi

CRC Press
Taylor & Francis Group
Boca Raton London New York

CRC Press is an imprint of the
Taylor & Francis Group, an **informa** business

CRC Press
Taylor & Francis Group
6000 Broken Sound Parkway NW, Suite 300
Boca Raton, FL 33487-2742

First issued in paperback 2023

© 2018 by Taylor & Francis Group, LLC
CRC Press is an imprint of Taylor & Francis Group, an Informa business

No claim to original U.S. Government works

ISBN-13: 978-0-8153-9291-0 (hbk)
ISBN-13: 978-1-03-265316-7 (pbk)
ISBN-13: 978-1-351-18951-4 (ebk)

DOI: 10.1201/9781351189514

Publisher's Note
The publisher has gone to great lengths to ensure the quality of this reprint but points out that some imperfections in the original copies may be apparent.

Library of Congress Cataloging-in-Publication Data

Names: Aggarwal, R. L. (Roshan Lal), 1937- author. | Alavi, Kambiz, author.
Title: Introduction to optical components / Roshan L. Aggarwal and Kambiz Alavi.
Description: First edition. | Boca Raton, FL : CRC Press, Taylor & Francis Group, 2018. | "A CRC title, part of the Taylor & Francis imprint, a member of the Taylor & Francis Group, the academic division of T&F Informa plc." | Includes bibliographical references and index.
Identifiers: LCCN 2017049572| ISBN 9780815392910 (hardback : acid-free paper) | ISBN 9781351189514 (ebook)
Subjects: LCSH: Optical instruments--Equipment and supplies. | Optical materials.
Classification: LCC TS513 .A44 2018 | DDC 620.1/1295--dc23
LC record available at https://lccn.loc.gov/2017049572

Visit the Taylor & Francis Web site at
http://www.taylorandfrancis.com

and the CRC Press Web site at
http://www.crcpress.com

Dedication

This book is dedicated to our parents (Chet Ram Aggarwal and Lila Vati Aggarwal, and Seyed Mohammed Kazem Alavi and Bibi Ozra Nadji Alavi), our spouses (Pushap Lata Aggarwal and Homa Rahmani-Khezri Alavi), and our children (Rajesh Aggarwal and Achal Aggarwal, and Maysa Alavi, Tara Alavi, and Kiana Alavi).

Contents

Preface

This book is intended to provide readers with a brief introduction to optical components. Material in this book will prepare readers for dealing with optical components in the area of optics and optical technology. The sources for the material in this book are several books on optics as well as information available from several vendors and others as acknowledged in the references at the end of the book. There are three outstanding features of this book: (1) It is relatively short; (2) The equations in this book are given without proof in order to keep the book short; (3) Numerous tables are given providing useful optical parameters for a large number of optical materials, making this book useful for optical design engineers. Additionally, this book includes an appendix with the solutions to homework problems. There are several books with optical components in their title. The most relevant are authored by Kai Chang: (1) *Handbook of Microwave and Optical Components, Volume 3: Optical Components* (1990); and (2) *Handbook of Optical Components and Engineering* (2003). However, there are no textbooks on optical components. The technical level of this book is equivalent to an undergraduate course in optics and optical technology curriculum. The students are required to have some familiarity with optics. Also, practitioners in optics and optical technology can use this book.

Acknowledgments

We thank Dr. Marc Bernstein for giving his approval to write this book. We thank Dr. William Herzog and Dr. Mordechai Rothschild for discussions regarding this work. We thank Dr. Antonio Sanchez-Rubio and Dr. Marion Reine for their comments on this work. We thank Peter O'Brien for providing data on high-reflection (HR) coatings, and cold and hot mirrors. We thank Dr. Alan DeCew for providing information regarding the Hubble Space Telescope. We thank *Applied Physics Letters*, Hamamatsu, *Nature*, RP Photonics, and Wiley-VCH for permission to reproduce Figures 8.2, 8.3, 8.5, 8.6, 9.4, and 9.5. We thank Edmund Optics, First Sensor, ISP Optics, Melles Griot, Newport Corporation, Nikon, Perkin-Elmer Corporation, Schott Advanced Optics, Semrock, Sustainable Supply, Thorlabs, Vision Tech Systems, and Wikipedia for the use of their online data. We thank Casey Reed and Tara Alavi for drafting many of the figures.

Authors

Roshan L. Aggarwal retired from Massachusetts Institute of Technology (MIT), Cambridge, Massachusetts, effective April 1, 2016 after 51 years of service. He is currently working as Part-Time Flexible Technical Staff in Chemical, Microsystem, and Nanoscale Technologies (Group 81) at MIT Lincoln Laboratory. Previously, he was Technical Staff at MIT Lincoln Laboratory for 30 years (1986–2016); Senior Research Scientist, MIT Physics Department for 12 years (1975–1987); Associate Director, MIT Francis Bitter National Magnet Laboratory for 7 years (1977–1984); and Technical Staff, MIT Francis Bitter National Magnet Laboratory for 12 years (1965–1977).

Kambiz Alavi is currently a Professor in the Electrical Engineering Department, University of Texas at Arlington, Arlington, Texas. He served 9 years as Associate Chairman (2008–2017). He was a Research Scientist at Siemens Corporate Research at Princeton, New Jersey (1983–1988) and a Postdoctoral Member of Technical Staff at AT&T Bell Laboratories in Murray Hill, New Jersey (1981–1983). He served as Site Director of NSF Industry/University Cooperative Research Center (CEMDAS) at UTA (1995–1997). During 2001–2003 he was a Department Manager in Integrated Optoelectronics, Advanced Systems and Technology, BAE Systems, Nashua, New Hampshire. He earned SB, SM, and PhD degrees in Physics from MIT. His research was conducted at MIT Francis Bitter National Magnet Laboratory.

1 Lenses

1.1 INTRODUCTION

The earliest known lenses date back to 750 BC, and were made from polished crystal, often quartz. One such example of an ancient lens in the collection of the British Museum (#90959) was ground and polished with one plane surface and one slightly convex surface. Lenses are commonly used as magnifiers for viewing small objects. Lenses are also used in optical systems such as binoculars, cameras, eyeglasses, microscopes, telescopes, and other optical systems. For additional information on lenses, the following books are recommended: Fowles (1975); Hecht and Zajac (1974); Jenkins and White (1976).

There are six types of lenses: biconvex, plano-convex, positive meniscus, negative meniscus, plano-concave, and biconcave, as shown in Figure 1.1.

We now consider a biconvex lens of refractive index n, diameter D, focal length f, front focal length f_F, and back focal length f_B located in a medium of refractive index of 1.0 (vacuum or air). Figure 1.2 shows a thick lens with radii of curvature R_1 and R_2 for the two lens surfaces, and center thickness t_C.

The optical axis of the lens (C_2C_1) intersects the two surfaces of the lens at points V_1 and V_2. V_1V_2 is equal to t_C. C_1 and C_2 denote the centers of curvature of the two lens surfaces. V_1C_1 is equal to R_1. V_2C_2 is equal to R_2. F_1 and F_2 denote the front and back focal points of the lens. H_1 and H_2 denote the primary and secondary principal planes, respectively. An object ray, propagating parallel to the optical axis to the right, is considered to continue through the lens up to the principal plane H_2 and then passes through the back focal point F_2. The distance F_1H_1 or H_2F_2 is equal to the focal length f. The distance F_1V_1 is equal to the front focal length f_F. The distance V_2F_2 is equal to the back focal length f_B. The distances V_1H_1 and V_2H_2 are given by (Fowles 1975).

$$V_1H_1 = \delta_1 = -ft_C\left(\frac{n-1}{nR_2}\right) \tag{1.1}$$

$$V_2H_2 = \delta_2 = -ft_C\left(\frac{n-1}{nR_1}\right) \tag{1.2}$$

Distances V_1C_1, V_2C_2, V_1H_1, V_2H_2, F_1H_1, and H_2F_2 are positive if they point to the right, and negative if they point to the left. As shown in Figure 1.2, R_1 is positive, R_2 is negative, δ_1 is positive, and δ_2 is negative. The edge thickness of the lens is given by

$$t_E \approx t_C - \frac{D^2}{8}\left(\frac{1}{R_1} - \frac{1}{R_2}\right) \tag{1.3}$$

1

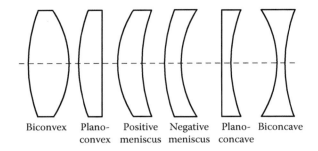

FIGURE 1.1   Six types of lenses.

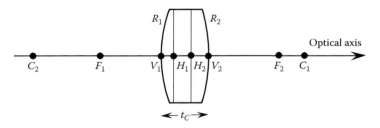

FIGURE 1.2   Thick lens with center thickness t_C.

In the paraxial approximation, focal length f is given by

$$\frac{1}{f} = (n-1)\left[\frac{1}{R_1} - \frac{1}{R_2} + \frac{(n-1)t_C}{nR_1R_2}\right] \tag{1.4}$$

Front focal length f_F is given by

$$f_F = f - \delta_1 \tag{1.5}$$

Back focal length f_B is given by

$$f_B = f + \delta_2 \tag{1.6}$$

Effective focal length of a combination of two thin lenses of focal lengths f_1 and f_2 separated by a distance d is given by

$$\frac{1}{f} = \frac{1}{f_1} + \frac{1}{f_2} - \frac{d}{f_1f_2} \tag{1.7}$$

Back focal length for this combination of the two lenses is given by

$$f_B = f\left(1 - \frac{d}{f_1}\right) \tag{1.8}$$

Location of the f lens on the left-hand side of the f_2 lens is given by

$$L_f^{f_2} = f\frac{d}{f_1} \tag{1.9}$$

Figure 1.3 shows a plane parallel beam of light incident upon a biconvex lens.

All the light rays pass through the focal spot and then diverge. The diameter of the focal spot is determined by the diffraction of light and the quality of the lens. For a perfect lens, the back focal spot F_2 consists of a bright spot known as the Airy disc, which is surrounded by rings. Eight-four percent of the light goes into the Airy disc and 16% into the rings. The diameter of the Airy disc is equal to $1.22f\lambda/D$, where λ is the wavelength of light and D is the diameter of the lens. For an $f/1.0$ perfect lens ($f\# = f/D$), the diameter of Airy disc is equal to 0.61 μm for λ equal to 0.5 μm. Because of the aberrations of the lens, the size of the focal spot for a real lens is larger than that for the perfect lens.

Figure 1.4 shows a plane parallel beam of light incident upon a biconcave lens. In this case, the light rays diverge upon exiting the lens and appear to be coming from a virtual focal spot, which is located on the left-hand side of the lens.

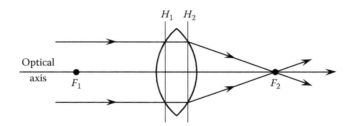

FIGURE 1.3 Plane parallel beam incident upon a biconvex lens.

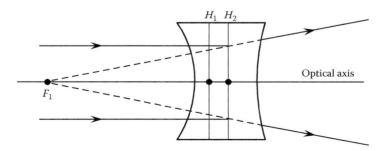

FIGURE 1.4 Plane parallel beam incident upon a biconcave lens.

1.2 MATERIALS

The optical properties of a material are determined by the values of its optical constants n and κ, where n is the refractive index and κ is the extinction coefficient.

The absorption coefficient α is given by

$$\alpha = \frac{4\pi\kappa}{\lambda} \tag{1.10}$$

A lens material should have negligible value of α ($<1 \times 10^{-6}/\mu m$). Table 1.1 lists wavelengths for negligible values of α for some materials.

Dispersion of n is given by the Sellmeier equation

$$n^2(\lambda) = 1 + \frac{B_1\lambda^2}{\lambda^2 - \lambda_1^2} + \frac{B_2\lambda^2}{\lambda^2 - \lambda_2^2} + \frac{B_3\lambda^2}{\lambda^2 - \lambda_3^2} \tag{1.11}$$

where B_1, B_2, B_3, λ_1^2, λ_2^2, and λ_3^2 are the Sellmeier coefficients. Equation 1.11 is not valid in regions where λ^2 is close to λ_1^2, λ_2^2, and λ_3^2. The Sellmeier coefficients are given in Table 1.2 for FS (Refractive Index 2017), BK7 (Refractive Index 2017), CaF$_2$ (Tatian 1984), BaF$_2$ (Tatian 1984), and Ge (Tatian 1984).

Values of n for FS, BK, CaF$_2$, and BaF$_2$ at 0.55 μm are 1.46, 1.52, 1.43, and 1.47 respectively. The value of n for Ge at 3.0 μm is 4.05.

Reflection loss R on each surface of a lens with negligible absorption is given by

$$R = \left(\frac{n-1}{n+1}\right)^2 \tag{1.12}$$

TABLE 1.1

Wavelengths for Negligible Values of α

Material	FS	BK7	CaF$_2$	BaF$_2$	Ge
Wavelengths	0.2–2.5 μm	0.4–2.5 μm	0.2–7 μm	0.2–9 μm	3–10 μm
Spectral region	UV, VIS, NIR	VIS, NIR	UV, VIS, NIR, IR	UV, VIS, NIR, IR	IR

Note: UV, VIS, NIR, and IR denote ultraviolet, visible, near-infrared, and infrared, respectively.

TABLE 1.2

Sellmeier Coefficients for FS, BK7, CaF$_2$, BaF$_2$, and Ge

Material	B_1	B_2	B_3	λ_1^2 (μm²)	λ_2^2 (μm²)	λ_3^2 (μm²)
FS	0.696166	0.407943	0.897479	0.004679	0.013512	97.93400
BK7	1.039612	0.231792	1.010470	0.006001	0.020018	103.5607
CaF$_2$	0.337601	0.701105	3.847815	0.000000	0.008775	1200.2
BaF$_2$	1.006307	0.143786	3.788478	0.000057	0.017520	2131.8
Ge	14.75875	0.235256	−24.8823	0.188619	1.593803	1695204

The values of R for FS, BK7, CaF$_2$, and BaF$_2$ at 0.55 μm are 3.5%, 4.3%, 4.3%, and 5.3%, respectively. The value of R for Ge at 3.0 μm is 36%. Lenses are antireflection (AR) coated to reduce the reflection loss. The AR coatings are usually broadband coatings. The AR coatings for specific wavelengths have a lower value of R than those for the broadband coatings.

1.3 IMAGING

Lenses are used for imaging applications. The image of an object, located at a distance s_o from a lens of focal length f, as shown in Figure 1.5, is obtained at a distance s_i given by

$$\frac{1}{s_i} = \frac{1}{f} - \frac{1}{s_o}$$

(1.13)

Object and image distances are measured from the principal planes H_1 and H_2, respectively. Image distance s_i is positive if the image is to the right of the lens. Magnification M of the image is given by

$$M = \frac{l_i}{l_o} = -\frac{s_i}{s_o}$$

(1.14)

where l_i and l_o are the lengths of the image and object, respectively. An image of an object formed by a lens, or combination of lenses, can be obtained by tracing the rays from the object through the lens (or lenses) using Snell's law. There are commercial software programs, such as Zemax, available for ray tracing.

1.4 ABERRATIONS

An object's image may not be a scaled replica of itself due to aberrations of the lens. There is chromatic aberration and five monochromatic aberrations, which are discussed in the following sections. Strehl ratio is a measure of the quality of the image. The Strehl ratio has a value between 0 and 1; an unaberrated image has a Strehl ratio of 1.

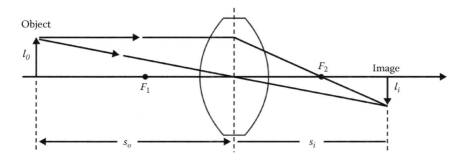

FIGURE 1.5   Image of an object formed by a lens.

1.4.1 CHROMATIC ABERRATION

Focal length of a lens depends on the wavelength λ due to the variation of n with λ. An image of an object with polychromatic light will consist of a series of images, one for each λ. This image defect is called chromatic aberration. The fractional change in the focal length over the spectral bandwidth $\Delta\lambda$ is given by

$$\frac{\Delta f}{f} = -\frac{\Delta n}{n-1} \qquad (1.15)$$

where Δn is the change in n over the spectral bandwidth $\Delta\lambda$. A combination of two lenses of different materials in contact with each other could result in a lens of zero chromatic aberration. Such a lens is called an achromatic doublet. The focal lengths of the two lenses of an achromatic doublet of focal length f are given by (Fowles 1975)

$$f_1 = f\left(1 - \frac{\Delta_1}{\Delta_2}\right) \qquad (1.16)$$

$$f_2 = f\left(1 - \frac{\Delta_2}{\Delta_1}\right) \qquad (1.17)$$

where

$$\Delta_1 = \frac{1}{(n_1 - 1)} \frac{dn_1}{d\lambda} \qquad (1.18)$$

$$\Delta_2 = \frac{1}{(n_2 - 1)} \frac{dn_2}{d\lambda} \qquad (1.19)$$

It is also possible to design an achromatic doublet using two lenses of the same material of focal lengths f_1 and f_2 separated by a distance d. It can be shown that an achromatic doublet is obtained if d is given by (Jenkins and White 1976)

$$d = \frac{1}{2}(f_1 + f_2) \qquad (1.20)$$

1.4.2 SPHERICAL ABERRATION

Focal length of a lens is a function of the distance h of the object rays from the optical axis. Normally, the focal length of a lens is specified for values of h, which are much smaller than the focal length, that is, in the paraxial approximation. For large values of h, the paraxial approximation is not valid. Consequently, the image of an on-axis object is degraded if the diameter of the lens is not much smaller than its focal length. This image defect is known as spherical aberration. The magnitude of the spherical aberration depends on the shape of the lens. The shape factor is given by (Jenkins and White 1976)

$$q = \frac{R_2 + R_1}{R_2 - R_1} \tag{1.21}$$

We use the third-order theory where

$$\sin\theta = \theta - \frac{\theta^3}{6} \tag{1.22}$$

Longitudinal spherical aberration (LSA) for a collimated ray incident upon the lens of focal length f at height h from the optical axis is given by

$$\text{LSA} \cong \frac{h^2}{8fn(n-1)}\left[\frac{n+2}{n-1}q^2 - 4(n+1)q + (3n+2)(n-1) + \frac{n^3}{n-1}\right] \tag{1.23}$$

Value of q for the minimum spherical aberration is obtained from Equation 1.23 as

$$q_{min} = \frac{2(n^2-1)}{n+2} \tag{1.24}$$

Value of q_{min} for $n = 1.5$ is 0.71, which is close to the value of 1.0 for q for a plano-convex lens with the first surface being convex. Spherical aberration may be further reduced using aspherized lens surfaces.

1.4.3 COMA

Coma derives its name from the pear-shaped comet-like appearance of the image of an off-axis point object located at a very large distance compared to the focal length of the lens. Assuming that the lens is free of spherical aberration, rays in the vicinity of the central ray form a sharp point image at A in the focal plane. On the other hand, rays from zones of increasing h form an image on a circle of increasing radius, as shown in Figure 1.6.

Radius of a comatic circle is given by (Jenkins and White 1976)

$$C_s = \frac{jh^3}{f^3}\left(\frac{3(2n+1)}{4n}p + \frac{3(n+1)}{4n(n-1)}q\right) \tag{1.25}$$

where:
j is the distance of the sharp point image A from the optical axis
h is the radius of the incident rays on the lens
p is the position factor defined by

$$p = \frac{2f}{s_o} - 1 = 1 - \frac{2f}{s_i} \tag{1.26}$$

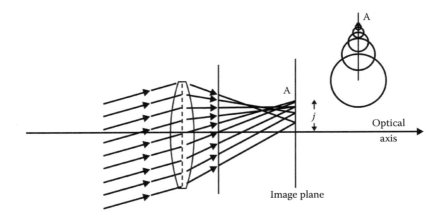

FIGURE 1.6   Coma-shaped image formed by a lens.

A lens will be free from comatic aberration for a distant off-axis point object, which corresponds to the value of −1 for p, if shape factor q is given by

$$q = \frac{(2n+1)(n-1)}{n+1} \tag{1.27}$$

Value of q is 0.80 for $n = 1.5$. This shows that a lens designed for no coma will have spherical aberration close to its minimum.

1.4.4   ASTIGMATISM

Astigmatism is a monochromatic third-order aberration, which arises when a point object lies an appreciable distance from the optical axis of the lens. An astigmatic lens produces two line images of an off-axis point object. One of these line images lies in the tangential plane, which contains the optical axis and the point object. The other line image lies in the sagittal plane, which is perpendicular to the tangential plane, as shown in Figure 1.7. The tangential image plane is closer to the lens than the sagittal image plane.

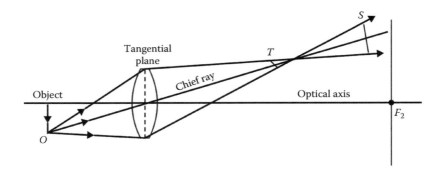

FIGURE 1.7   Two line images of an off-axis point object formed by an astigmatic lens.

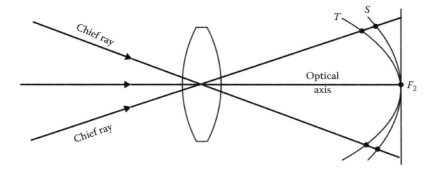

FIGURE 1.8   Tangential and sagittal image planes for an off-axis point object.

Figure 1.8 shows that the separation between the tangential and sagittal image planes increases with the distance of the point object from the optical axis. The two image surfaces coincide for paraxial rays. The astigmatism is approximately proportional to the focal length of the lens as given below:

$$(s_i)_S - (s_i)_T \approx f \sin^2 \phi \tag{1.28}$$

where:

$(s_i)_S$ and $(s_i)_T$ are the image distances in the sagittal and tangential planes, respectively, and

ϕ is the angle of incidence of the chief ray measured at the point of incidence on the lens.

Astigmatism does not depend upon the shape of the lens. This is in contrast to spherical aberration and coma, both of which depend upon the shape of the lens. A second form of astigmatism occurs when the optical system is not symmetric about the optical axis. This may be due to manufacturing errors in the surfaces of the lens.

Astigmatism is well illustrated by looking at the image of a spoked wheel with its center located on the optical axis and in a plane perpendicular to the optical axis (parallel to the plane of the lens), as shown in Figure 1.9 (Jenkins and White 1976).

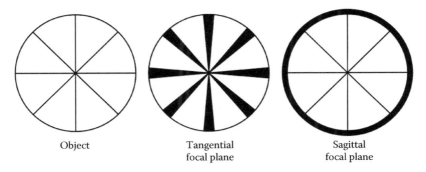

Object   Tangential   Sagittal
                          focal plane   focal plane

FIGURE 1.9   Astigmatic images of a spoked wheel formed by a lens.

Rim is in focus in the tangential T image plane while spokes are in focus in the sagittal image plane p. S. An off-axis point object is imaged in the sagittal focus as a line parallel to the spokes. An off-axis point object is imaged in the tangential plane as a line parallel to the rim. That is why the spokes of the wheel are in focus in the sagittal plane and the rim is in focus in the tangential plane.

1.4.5 FIELD CURVATURE

Field curvature is an aberration that is related to astigmatism, but it can also exist in a system that does not suffer from astigmatism. An optical system or a lens free of spherical aberration, coma, and astigmatism will provide a one-to-one correspondence between points on the object and image surfaces. However, a planar object normal to the optical axis is imaged approximately as a plane only in the paraxial approximation. In general, the image surface would be curved. This image defect is known as Petzval field curvature after the Hungarian mathematician Joseph Max Petzval. The Petzval surface for a single lens is given by

$$\Delta z = \frac{x^2}{2nf} \tag{1.29}$$

where:
   z is the coordinate along the optical axis of the lens
   x is the coordinate in the image plane

The Petzval surface does not depend upon the shape of the lens. For a two-lens system, the Petzval surface can be made planar if

$$n_1 f_1 + n_2 f_2 = 0 \tag{1.30}$$

where:
   n_1 and n_2 are the refractive indices
   f_1 and f_2 are the focal lengths of the two lenses

Let us consider the case of two lenses separated by a distance d. If $n_1 = n_2$ and $f_1 = -f_2$ then Equation 1.30 is valid. In this case, the Petzval surface for the combination of the two lenses will be planar and the combination will have a finite positive focal length given by

$$f = \frac{f_1^2}{d} \tag{1.31}$$

1.4.6 DISTORTION

In the absence of spherical aberration, coma, astigmatism, and field curvature, each point on a planar object would be sharply focused on a planar image plane. However, the image would be distorted if the transverse magnification M_T were not uniform over the entire field of view. There are two forms of distortion: (1) barrel distortion, which results when the magnification decreases toward the edges of the field of view; and (2) pincushion distortion, which results when the magnification increases toward

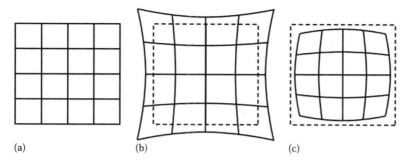

FIGURE 1.10   (a) Object, (b) pincushion distortion, and (c) barrel distortion.

the edges of the field of view. The increase (decrease) of magnification for pincushion (barrel) distortion varies quadratically with radial distance. These two forms of distortion are shown in Figure 1.10 for a wire screen.

1.5   MAGNIFIER

A magnifier is a positive lens, which provides a magnified image of an object on the retina of the eye rather than that obtained with the unaided eye. The angular magnification of the magnifier is the ratio of the angle θ' subtended by the image to the angle θ subtended by the object. The magnification of the magnifier is specified by (Jenkins and White 1976)

$$M = \frac{\theta'}{\theta} = \frac{25}{f} + 1 \approx \frac{25}{f} \tag{1.32}$$

This is illustrated in Figure 1.11.

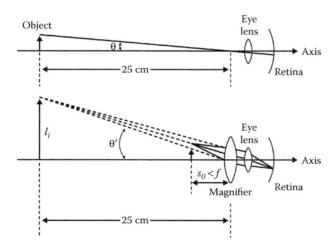

FIGURE 1.11   Angles θ and θ' subtended by the object at the unaided eye and that of the image of the magnifier.

Here 25 cm is the standard near point, which is called the distance of most distinct vision, and f is the focal length of the magnifier in units of cm. A magnifier with a focal length of 2.5 cm is marked 10x and another with a focal length of 5 cm will be marked 5x.

1.6 OBJECTIVES

Objective lenses are used to collect light from the object and then focus the light to form a real image. Objective lenses are used in cameras, microscopes, and telescopes. The microscope objective has a very short focal length. The magnification of the microscope objective typically ranges from 4x to 100x. The telescope objective has a relatively large focal length (several inches to tens of inches). The light collection of the telescope objective is proportional to the square of the diameter of the objective. The world's largest refracting telescope is located at the Yerkes Observatory in Wisconsin Bay, Wisconsin, US. The diameter and focal length of the Yerkes objective are 1.02 m and 19.4 m, respectively.

1.7 EYEPIECES

An eyepiece is used to magnify the image created by the objective lens in a microscope, telescope, or other optical devices such as binoculars. The eyepiece is so named because it is close to the eye in the optical device. The eyepiece can be a single lens, but usually consists of several lenses to reduce the aberrations of the image. Also, the component lenses of the eyepiece are AR-coated to improve its transmittance. The angular magnification of the eyepiece is specified in the same way as for the magnifier. The overall magnification of a microscope M_M is a product of the linear magnification M_O of the objective and angular magnification of the eyepiece M_E. Hence M_M is given by

$$M_M = M_O M_E \tag{1.33}$$

Assuming values of 100 for M_O and 10 for M_E, the value of M_M will be equal to 1000, which allows the observation of submicron-sized objects. The magnification M_T of a telescope is defined as the ratio of the angle subtended at the eye by the final image and the angle subtended at the eye by the object itself. M_T is given by

$$M_T = \frac{f_O}{f_E} \tag{1.34}$$

where f_O and f_E are the focal lengths of the telescope objective and eyepiece, respectively. A telescope with an objective of 5 m focal length and an eyepiece of 5 cm focal length will have a magnification of 100.

1.8   CAMERA LENSES

Camera lenses have either fixed focal length or zoom lenses that have adjustable focal lengths. The dedicated digital camera lenses are required to cover a relatively large area in the focal plane of the lens. Therefore, camera lenses have numerous lens elements, including aspheric elements to correct for the chromatic and monochromatic aberrations. Depth of field (DOF) of a camera lens is the distance between the farthest and nearest objects in the scene that appear sharp in the image. DOF is given by

$$\text{DOF} \approx 2(f\#)\delta\left(\frac{s_o}{f}\right)^2 \qquad (1.35)$$

where:
  $f\#$ is the f-number of the lens
  δ is the diameter of the circle of confusion that is indistinguishable from a point
  s_o is the object distance for sharp focus
  f is the focal length of the lens

For example, DOF has a value of 0.64 m using values of 2.0, 25 μm, 4.0 m, and 50 mm for $f\#$, δ, s_o, and f, respectively. Field of view is inversely proportional to the focal length of the lens as shown in Figure 1.12.

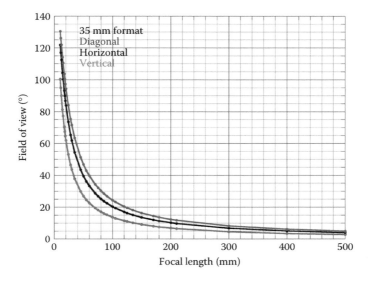

FIGURE 1.12   Field of view of a camera lens. (Courtesy of Nikon, www.nikonins.org, 2017.)

1.8.1 FIXED FOCAL LENGTH CAMERA LENSES

Typical fixed focal lengths of the camera lenses are 35, 50, and 85 mm. Typical *f*-numbers are 2.8, 1.8, 1.4, and 1.2. In photography, 50 mm is considered to be the *normal* focal length. Whenever you look through a 50-mm lens, the scene appears to be the same as seen with your own eye. A longer focal length lens magnifies the image by a factor, which is equal to the focal length divided by 50. For example, a 200-mm lens magnifies everything in the scene by a factor of 4. An 85-mm lens is referred to as the *portrait* lens. A 35-mm lens is referred to as the *street* lens. The longer the focal length, the shallower is the depth of field. A 100-mm lens at *f*/2.8 will have a much shallower depth of field than a 35-mm lens at *f*/2.8.

1.8.2 ZOOM LENSES

Typical zoom lenses cover 24–70, 70–300, or 28–200 mm focal lengths. A zoom lens is an assembly of lens elements for which the focal length can be varied. A simple design of a zoom lens consists of two parts: an afocal zoom system preceding a focusing lens, as shown in Figure 1.13. The afocal zoom system consists of two lenses (positive) L_1 and L_3 of equal focal length, and a negative lens L_2 with the magnitude of the focal length less than half that of the positive lenses. The negative lens is placed between the two positive lenses. The first positive lens L_1 and the negative lens L_2 are moved axially in a non-linear fashion. The afocal zoom system does not focus light but alters the size of the beam of light traveling through it. The short focal length is obtained with L_2 close to L_1, and the long focal length is obtained with L_2 close to L_3, as shown in the homework problem 1.18.

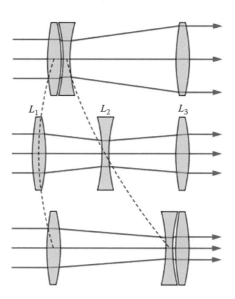

FIGURE 1.13 Afocal zoom lens system. (Courtesy of Wikipedia, www.wikipedia.org, 2017.)

Wide-angle zoom lenses (marked W) cover focal lengths under 50 mm. At any focal length under 50 mm, the lens brings more of the surroundings into the frame.

1.8.3 Telephoto Lenses

A telephoto lens is a long-focus lens in which the physical length of the lens is shorter than the focal length. The telephoto lens consists of a large positive lens group and a smaller negative lens group. The positive group has a much shorter focal length than that of the equivalent long-focus lens. The telephoto lens has a narrow field of view. A medium telephoto lens has a focal length greater than 70 mm, and a full telephoto lens has a focal length greater than 135 mm.

1.8.4 Cell Phone Camera Lenses

Worldwide, more people use cell phone cameras now than the dedicated digital cameras. The principal advantages of a cell phone camera are its cost and compactness. The patents for the cell phone camera date back as far as 1956. On June 11, 1997, Philippe Kahn wirelessly sent the first pictures of his daughter, Sophie, from the maternity ward to more than 2000 family, friends, and associates around the world. Kahn's cell phone transmission is the first known publicly shared picture via a cell phone. The first cell phone cameras were sold by J-Phone in Japan in 2000. Cell phone camera lenses have a short (4–5 mm) fixed focal length in order to fit within the size of the cell phone.

1.9 OTHER LENSES

The common lenses discussed above are based on homogeneous lens material. These lenses suffer from spherical and other aberrations. There are other types of lenses that are used to minimize aberrations, or to reduce weight and size. These include aspherical lenses, gradient-index (GRIN) lenses, diffractive optical elements, photonic crystals, and wavefront adaptive lenses.

Graded index lenses use compositional grading to achieve an index of refraction that varies smoothly inside the lens. In the graded index lens, optical rays undergo gradual bending towards a focus. The lens of the eye is the most obvious example of gradient-index optics in nature. In the human eye, the refractive index of the lens varies from approximately 1.406 in the central layers down to 1.386 in less dense layers of the lens (Hecht and Zajac 1974).

1.10 HOMEWORK PROBLEMS

1.1 A biconvex lens has refractive index of 1.5, 25 mm diameter, 10 mm center thickness t_C, and radii of curvature R_1 and R_2 of 5.0 and −10.0 cm, respectively. Calculate f, δ_1, δ_2, f_F, f_B, and the edge thickness t_E. Make a sketch showing the lens surfaces, optical axis, principal planes, and points V_1, V_2, H_1, H_2, F_1, and F_2.

1.2 A biconcave lens has refractive index of 1.5, 25 mm diameter, 10 mm center thickness t_C, and radii of curvature R_1 and R_2 of −5.0 and 10.0 cm, respectively. Calculate f, δ_1, δ_2, f_F, f_B, and the edge thickness t_E. Make a sketch showing the lens surfaces, optical axis, principal planes, and points V_1, V_2, H_1, H_2, F_1, and F_2.

1.3 A positive meniscus lens has refractive index of 1.5, 25 mm diameter, 10 mm center thickness t_C, and radii of curvature R_1 and R_2 of 2.5 and 5.0 cm, respectively. Calculate f, δ_1, δ_2, f_F, f_B, and the edge thickness t_E. Make a sketch showing the lens surfaces, optical axis, principal planes, and points V_1, V_2, H_1, H_2, F_1, and F_2.

1.4 A negative meniscus lens has refractive index of 1.5, 25 mm diameter, 10 mm center thickness t_C, and radii of curvature R_1 and R_2 of 5.0 and 2.5 cm, respectively. Calculate f, δ_1, δ_2, f_F, f_B, and the edge thickness t_E. Make a sketch showing the lens surfaces, optical axis, principal planes, and points V_1, V_2, H_1, H_2, F_1, and F_2.

1.5 A 10x He-Ne laser beam expander consists of two lenses of focal lengths f_1 and f_2. The value of f_1 is 10 mm. Calculate the values of f_2, and distance d between the two lenses. The diameter of the input beam D_i is 2 mm. Calculate the diameter of the output beam D_o.

1.6 A 10x He-Ne laser beam expander consists of two lenses of focal lengths f_1 and f_2. The value of f_1 is −10 mm. Calculate the values of f_2 and distance d between the two lenses. The diameter of the input beam D_i is 2 mm. Calculate the diameter of the output beam D_o.

1.7 An on-axis point object is located at a distance of 10.0 cm from a lens of 5.0 cm focal length. Determine the distance of the image of the object from the lens. Also, determine the magnification M of the image.

1.8 An on-axis point object is located at a distance of 20.0 cm from a lens of 5.0 cm focal length. Determine the distance of the image from the lens. Also, determine the magnification M of the image point.

1.9 The 25 and 40 mm focal length lenses are separated by 30 mm. A 3-mm long object is located at 30 mm on the left side of the 25-mm focal length lens. Calculate the size and location of the image of the object relative to that of the 40-mm focal length lens.

1.10 Calculate and plot the refractive index n of fused silica (FS), BK7, calcium fluoride (CaF_2), barium fluoride (BaF_2), and germanium (Ge) using Sellmeier equation over the wavelength regions 0.3–2.5, 0.3–2.5, 0.3–10, 0.3–10, and 3–10 µm, respectively.

1.11 A BK7 glass lens has 25.000 mm focal length at 0.55 µm. Calculate the change in its focal length over the spectral region 0.40–0.7 µm.

1.12 Calculate the longitudinal spherical aberration at 0.55 µm for an $f/1.8$ BK7 plano-convex lens of 47.2 mm diameter and 85.0 mm focal length. Assume that the plane parallel beam is incident upon the convex surface of the lens.

1.13 Calculate the maximum radius of the comatic circle for a distant, off-axis point object, which forms a sharp point image at a distance of 5 mm for the chief ray, assuming that the lens is plano-convex of refractive index 1.5 with 25 mm diameter and 25 mm focal length.

1.14 Calculate the shape factor of a coma-free germanium lens at 6-µm wavelength for a distant off-axis point object.

1.15 Calculate the separation between the image distances in the sagittal and tangential planes if the focal length of the lens is 50 mm and the angle of incidence of the chief ray at the lens is 10°.

1.16 Two lenses of the same refractive index with focal lengths $f_1 = 30$ mm and f_2 are separated by a distance of 30 mm. Determine the focal length of the combination of these lenses if the Petzval surface of the combination is planar.

1.17 Determine the magnification M_T of a telescope that consists of an objective of 2 m focal length and an eyepiece of 5 cm focal length.

1.18 Determine the depth of focus for a close-up object at 25 cm from a 50 mm focal length, $f/1.8$ camera lens, assuming a value of 25 µm for diameter of the circle of confusion.

1.19 A zoom lens consists of a 100-mm focusing lens L_0 together with an afocal zoom lens system that contains three lenses L_1, L_2, and L_3, as shown in Figure 1.13. The focal lengths of the L_1, L_2, and L_3 lenses are 200, −62.5, and 200 mm, respectively. The separation between L_1 and L_3 is 120 mm. The separation between L_3 and L_0 is negligible. Determine the effective focal length and back focal length of the zoom lens when (1) L_2 is in contact with L_1, and (2) L_2 is in contact with L_3, assuming that all four lenses are thin.

1.20 A zoom lens consists of three thin lenses L_1, L_2, and L_3, which have focal lengths of 200, −62.5, and 200 mm, respectively. The L_2 lens is moved from L_1 to L_3. The separation between L_1 and L_3 is 120 mm. Calculate the effective focal length and back focal length of the zoom lens when the separation between L_1 and L_2 is equal to 0, 10, 20, 30, 40, 50, 60, 70, 80, 90, 100, 110, and 120 mm.

1.21 An object is located at a distance of 70 mm in front of a relay telescope, which consists of two lenses of focal lengths f_1 equal to 30 mm and f_2 equal to 15 mm. The separation between the two lenses is equal to 45 mm ($f_1 + f_2$). The object is 5 mm in length. Determine (1) the length of the image, and (2) the location of the image formed by the relay telescope.

2 Mirrors

2.1 INTRODUCTION

The earliest man-made mirrors were of polished stone, some of which have been found in Turkey and date back at least 6000 years (http://www.mirrorhistory.com). Ancient Egyptians used a sheet of polished copper as a mirror. Back-surface mirrors made of metal-backed glass were first produced in Lebanon in the first century AD. Justus von Liebig, German chemist, produced the silver-coated glass mirror in 1835 through the deposition of a thin layer of silver by the chemical reduction of silver nitrate. This technique made the mass production of mirrors possible.

Most front-surface modern mirrors are made by the deposition of a thin layer of aluminum on glass under vacuum. A $\lambda/2$ coating of silicon monoxide (SiO) is typically used as an overcoat for the protection of the front surface aluminum layer. A multi-layer dielectric film on top of aluminum is used to enhance the reflectance in the VIS or UV regions. Figure 2.1 shows the reflectance spectrum of UV enhanced aluminum for the 200–1200 nm spectral range, which covers UV, VIS, and NIR.

Figure 2.2 shows the reflectance spectrum of protected silver for the spectral region 0.5–10 µm, which covers part of VIS, NIR, and IR. The reflectance of protected silver is greater than 97% over this spectral range.

Figure 2.3 shows the reflectance spectrum of protected gold for the spectral region 1–10 µm, which covers NIR and IR. The reflectance of gold is greater than 97%.

Typical laser energy density limits are 0.5 J/cm^2 (10 ns pulses) at 532 nm and 1064 nm for UV enhanced aluminum and protected silver, and 0.8 J/cm^2 (10 ns pulses) at 1064 nm for protected gold (Edmund Optics 2017).

Dielectric mirror coatings can be used for laser applications that require ultra-high reflectance (>99%) at a specific design wavelength. Figure 2.4 shows a 21-layer dielectric stack for producing high reflectance (HR). The HR stack consists of alternate quarter-wave (inside the material) layers of high and low refractive index materials. These multilayer layers are referred to as distributed Bragg reflectors (DBR). In semiconductor, vertical-cavity surface emitting diode lasers (VCSEL), the gain layer is sandwiched between two DBR mirrors, which are usually made of alternating semiconductor layers with differing indices of refraction.

Reflectance of a 2N-layer stack is given by (Fowles 1975)

$$R = \left[\frac{(n_H/n_L)^{2N} - 1}{(n_H/n_L)^{2N} + 1} \right]^2 \tag{2.1}$$

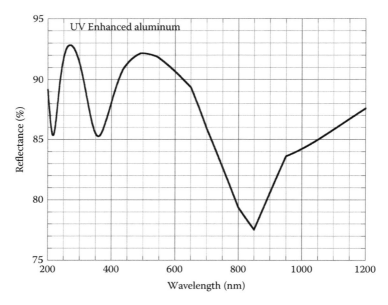

FIGURE 2.1   Reflectance UV enhanced aluminum. (Courtesy of Edmund Optics, www. edmundoptics.com/optics, 2016.)

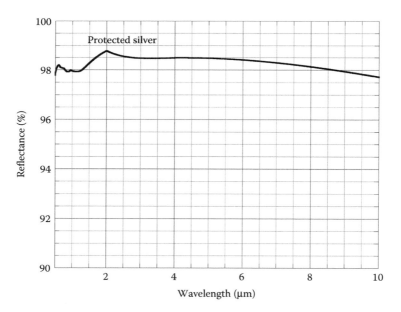

FIGURE 2.2   Reflectance of protected silver. (Courtesy of Edmund Optics, www. edmundoptics.com/optics, 2016.)

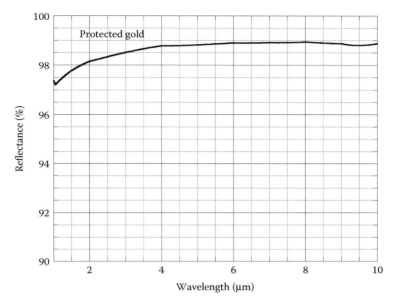

FIGURE 2.3 Reflectance of protected gold. (Courtesy of Edmund Optics, www. edmundoptics.com/optics, 2016.)

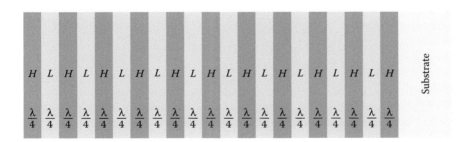

FIGURE 2.4 21-layer dielectric stack for a high reflectance film; λ is the wavelength of light in the layer.

Figure 2.5 shows the reflectance spectra at 0° and 45° angles of incidence (AOI) for the 21-layer ¼ wave HR stack of TiO_2 and SiO_2 on fused silica substrate. The refractive indices n of TiO_2 and SiO_2 are 2.55 and 1.50, respectively. The design wavelength for the HR stack is 532 nm. The layer thicknesses of TiO_2 and SiO_2 are 532/4n and 577/4n nm for the 0° and 45° AOI, respectively. The reflectivity is > 0.9995% in the 490–580 nm range.

Dielectric mirror coatings can withstand moderate to high laser powers. Laser damage threshold (LDT) measured in J/cm^2 for a laser mirror is dependent on the

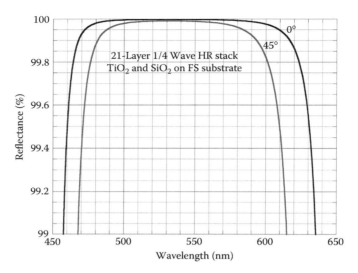

FIGURE 2.5   Reflectance of a 21-layer ¼ wave HR stack on fused silica. (Courtesy of Peter O'Brien.)

laser wavelength λ, pulse length τ, and rep rate ν. If LDT (λ_1, τ_1) is known for λ_1 and τ_1 in the 1–100 ns regime, LDT for λ_2 and τ_2 can be estimated from the following equation:

$$\text{LDT}(\lambda_2, \tau_2) \approx \text{LDT}(\lambda_1, \tau_1) \left(\frac{\lambda_2}{\lambda_1} \right) \sqrt{\frac{\tau_2}{\tau_1}} \tag{2.2}$$

Mirrors are available in a variety of substrate materials including fused silica, BK7, silicon, aluminum, and copper. For additional information, the following books are recommended: Jenkins and White (1976) and Hecht and Zajac (1974).

2.2   PLANE MIRRORS

Plane mirrors are either back or front surfaced. The back-surfaced mirrors are commonly used for inspection, as well as for everyday use, because the metallic reflecting layer is protected behind the glass. The front-surface mirrors are used for scientific and technological applications. Figure 2.6 shows the law of reflection for mirrors. The angle of reflection θ_r is equal to the angle of incidence θ_i. The angles θ_r and θ_i are measured from the normal to the plane mirror.

A plane mirror is commonly used to change the propagation direction of a light beam. Figure 2.7 shows that if the mirror is rotated through an angle α, then the reflected ray rotates through 2α.

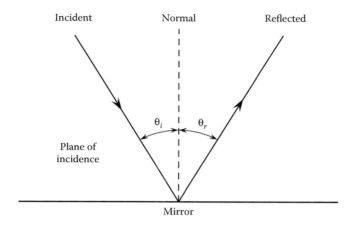

FIGURE 2.6 Angles of incidence and reflection at the surface of a mirror.

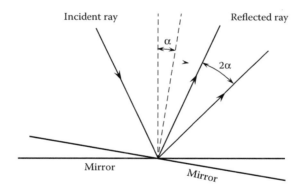

FIGURE 2.7 Schematic showing that the reflected ray rotates through an angle 2α when the mirror is rotated through an angle α.

Figure 2.8 shows the image of an object formed by a mirror. The object of length L_o is located at a distance s_o from the mirror. A virtual image of length L_i equal to L_o is located behind the mirror at a distance of s_i equal to s_o.

Cold mirrors are highly reflective at wavelengths below a certain value. They are highly transparent at longer wavelengths. This is the optical analog of a low-pass filter in electronics. Cold mirrors have a dielectric coating, which has high reflectance for visible light and low reflectance for IR. Figure 2.9 shows reflectance of a cold mirror with a 24-layer dielectric coating for use at 0° angle of incidence.

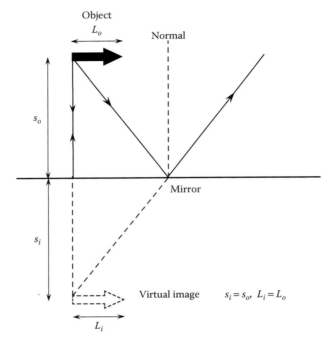

FIGURE 2.8   Image of an object formed by a mirror.

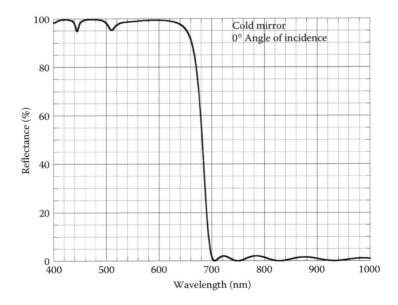

FIGURE 2.9   Reflectance of a 0° cold mirror. (Courtesy of Peter O'Brien.)

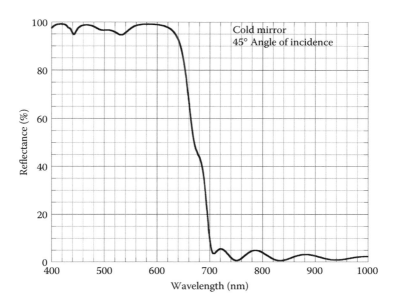

FIGURE 2.10 Reflectance of a 45° cold mirror. (Courtesy of Peter O'Brien.)

The structure of the dielectric coating in Figure 2.9 is substrate, 33.73 nm (*H*), 53.39 nm (*L*), 47.87 nm (*H*), 44.4 nm (*L*), 38.02 nm (*H*), 88.77 nm (*L*), 47.46 nm (*H*), 72.53 nm (*L*), 42.4 nm (*H*), 50.9 nm (*L*), 22.56 nm (*H*), 88.51 nm (*L*), 48.6 nm (*H*), 86.09 nm (*L*), 53.59 nm (*H*), 98.55 nm (*L*), 57.1 nm (*H*), 90.53 nm (*L*), 50.44 nm (*H*), 90.75 nm (*L*), 58.43 nm (*H*), 113.09 nm (*L*), 17.88 nm (*H*), 4.3 nm (*L*), air—where *H* denotes the high refractive index (2.55) layer of TiO_2, *L* denotes the low refractive index (1.50) layer of SiO_2, and the values denote the physical thickness of the layers.

Figure 2.10 shows the average of the s and p polarization reflectance spectrum of a cold mirror for use at 45°.

The structure of the dielectric coating in Figure 2.10 is substrate, 41.22 nm (*H*), 56.61 nm (*L*), 53.04 nm (*H*), 40.82 nm (*L*), 34.08 nm (*H*), 100.29 nm (*L*), 51.77 nm (*H*), 74.45 nm (*L*), 56.52 nm (*H*), 73.03 nm (*L*), 19.68 nm (*H*), 86.7 nm (*L*), 43.04 nm (*H*), 92.6 nm (*L*), 57.57 nm (*H*), 106.99 nm (*L*), 58.71 nm (*H*), 97.41 nm (*L*), 53.83 nm (*H*), 101.98 nm (*L*), 61.46 nm (*H*), 119.53 nm (*L*), 23.89 nm (*H*), air.

Hot mirrors use a multi-layer dielectric coating, which has low reflectance for visible light and high reflectance for IR. Figure 2.11 shows the reflectance spectrum of a hot mirror for use at 0° angle of incidence.

Figure 2.12 shows the average of the s and p polarization reflectance spectrum of a hot mirror for use at 45°.

The structure of the dielectric coating in Figure 2.12 is substrate, 90.98 nm (*H*), 161.83 nm (*L*), 85.14 nm (*H*), 147.71 nm (*L*), 81.20 nm (*H*), 150.62 nm (*L*), 82.77 nm (*H*), 152.56 nm (*L*), 101.06 nm (*H*), 29.72 nm (*L*), 102.02 nm (*H*), 157.60 nm (*L*), 86.02 nm (*H*), 164.62 nm (*L*), 96.56 nm (*H*), 201.10 nm (*L*), 108.38 nm (*H*), 202.23 nm (*L*), 108.24 nm (*H*), 190.20 nm (*L*), 90.99 nm (*H*), 168.26 nm (*L*), 82.89 nm (*H*), 71.77 (*L*), air.

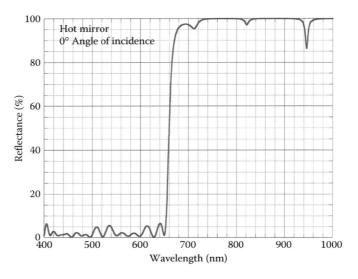

FIGURE 2.11   Reflectance of a 0° hot mirror. (Courtesy of Peter O'Brien.)

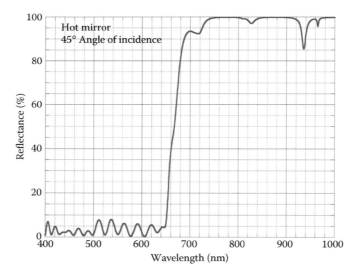

FIGURE 2.12   Reflectance of a 45° hot mirror. (Courtesy of Peter O'Brien.)

2.3 SPHERICAL MIRRORS

The focal length f of a concave (convex) spherical mirror is equal to ½ the radius of the curvature R of the spherical mirror surface. The sign convention is that R and f are positive for a concave mirror and negative for a convex mirror. Figure 2.13 shows a concave spherical mirror with its vertex at point V. The center of curvature is at point C.

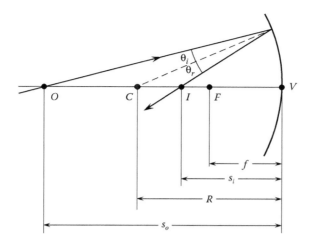

FIGURE 2.13 Concave spherical mirror.

The focal point is at F. The mirror produces an image of an on-axis point object. The object is located at point O. The image is located at point I. The object distance is OV, which is equal to s_o. The image distance is IV, which is equal to s_i. s_o and s_i are positive if the object and image are in front of the mirror.

Object and image distances s_o and s_i are related by

$$\frac{1}{s_o} + \frac{1}{s_i} = \frac{1}{f}$$

(2.3)

If the image distance s_i is equal to f, then s_o is equal to infinity. When the image distance s_i is equal to infinity, then s_o is equal to f, which is a collimated beam of light obtained for the object located at the focal point. Notably, mirrors are free of chromatic aberration. If the object and image are of length l_o and l_i, respectively, the magnification M of the image is given by

$$M = -\frac{l_i}{l_o} = -\frac{s_i}{s_o}$$

(2.4)

The negative sign on the RHS of Equation 2.4 implies the image is upside down.

Figure 2.14 shows a convex spherical mirror with its center of curvature at point C and virtual focal point at F'. A collimated beam of light incident upon the convex mirror is reflected as a divergent beam that appears to originate from F'. Equation 2.3 is also valid for the convex mirror provided f is taken to be negative. Then s_i is equal to f because s_o is equal to infinity, which corresponds to a plane parallel beam incident upon the convex mirror.

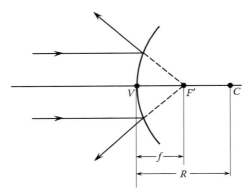

FIGURE 2.14 Convex spherical mirror with a collimated beam of incident light.

Convex spherical mirrors ($f < 0$) are used as side-view mirrors for cars. The virtual image of an object is always smaller than the object because the magnitude of the magnification is given by

$$M = \frac{|f|}{|f| + s_o} \tag{2.5}$$

M is given by $0 < M < 1$ for all values of the object distance s_o. This leads one to infer that the actual object is also smaller and hence farther away. That is why car mirrors have a warning sign "objects in mirror are closer than they appear."

2.4 PARABOLIC MIRRORS

The surface of a parabolic mirror is generated by rotation of a parabola around its axis. Figure 2.15 shows sections of a parabolic and concave spherical mirror of radius of curvature R.

The equation for the parabola is

$$x = \frac{y^2}{2R} \tag{2.6}$$

The equation for the circular arc is

$$x = \frac{y^2}{2R} + \frac{y^4}{8R^3} + \frac{y^6}{16R^5} + \cdots \tag{2.7}$$

The first term on the RHS of Equation 2.7 corresponds to that of a parabola. The circular arc and the parabola are indistinguishable for small values of y. The focal length f of the parabolic mirror is equal to $R/2$, same as that of the spherical mirror.

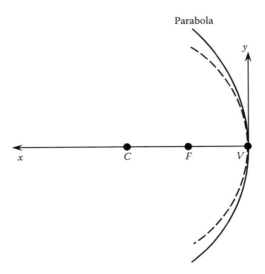

FIGURE 2.15   Comparison of concave spherical mirror with parabolic mirror.

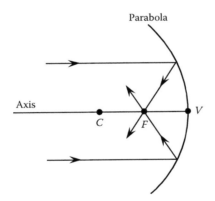

FIGURE 2.16   On-axis concave parabolic mirror.

Figure 2.16 shows an on-axis concave parabolic mirror. A plane parallel beam of light incident upon this mirror is focused at focal point F. The parabolic mirror is free of spherical aberration for infinitely distant on-axis objects. On-axis parabolic mirrors are used as collimators for light sources located at the focal point. Large-diameter parabolic mirrors are used for light collection applications such as telescopes.

Figure 2.17 shows an off-axis parabolic mirror, which is commonly used in optical spectrometers for collimating light focused on the entrance slit, which is located at the focus of the parabolic mirror. The off-axis parabolic mirror is also used to focus the collimated beam of light on the exit slit of the spectrometer.

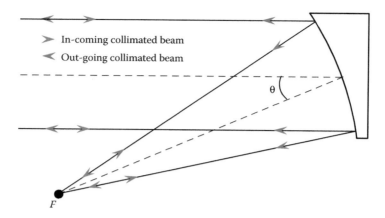

FIGURE 2.17 Off-axis concave parabolic mirror.

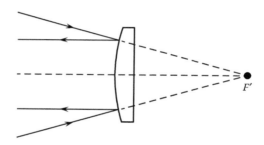

FIGURE 2.18 Convex parabolic mirror.

Figure 2.18 shows a convex parabolic mirror, which is used to collimate a beam of light that is converging to its virtual focal point.

2.5 HYPERBOLIC MIRRORS

The equation of a hyperbola is

$$\frac{(x+a)^2}{a^2} - \frac{y^2}{b^2} = 1 \tag{2.8}$$

The vertex of the mirror is the origin of the x-y coordinate system. The focal length of a hyperbolic mirror is given by

$$f = \frac{b^2}{2a} \tag{2.9}$$

The conic constant of the hyperbola is given by

$$K = -\left(1 + \frac{b^2}{a^2}\right) \tag{2.10}$$

For given values of f and K, one can determine the values of a and b. The Hubble Space Telescope (HST) exemplifies a Cassegrain telescope with a Ritchey-Chrétien design where the primary and secondary mirrors are both hyperbolic mirrors. The diameter and focal length of the primary mirror of the HST are 2.4 m and 5.52 m, respectively. The values of a and b for the HST primary mirror are 4803.13 m and 230.28 m, respectively. The diameter, focal length, and conic constant of the secondary mirror of the HST are 0.282 m, −0.679 m, and −1.49686 m, respectively. The values of a and b for the HST secondary mirror are 2.73 m and 1.93 m, respectively.

The focal length f of a combination of two mirrors of focal lengths f_1 and f_2 and separated by distance d is given by

$$\frac{1}{f} = \frac{1}{f_1} + \frac{1}{f_2} - \frac{d}{f_1 f_2} \tag{2.11}$$

or

$$f = \frac{f_1 f_2}{f_1 + f_2 - d} \tag{2.12}$$

The value of f for the HST is 57.6 m. Using values of 5.52 m for f_1 and −0.679 m for f_2, Equation 2.12 yields a value of 4.91 m for the mirror separation d. Back focal length, the distance of the focal spot s_i relative to the vertex of the secondary mirror of the HST, is equal to 6.36 m.

HST operates in the near UV, VIS, and NIR. HST is able to take high-resolution pictures because its orbit, which is 541.8 km above Earth's surface, lies outside the distortion of the Earth's atmosphere. The mirror of the HST was polished to an accuracy of 10 nm to take full advantage of the space environment. HST was launched into low Earth orbit on April 24, 1990 and entered service on May 20, 1990.

The Subaru Telescope of the National Astronomical Observatory of Japan (NAOJ), atop Mauna Kea in Hawaii, has a 8.3-m diameter primary hyperbolic mirror—one of the world's largest optical mirrors, which was polished by Contraves Brashear Systems (CBS) in the suburbs of Pittsburgh, Pennsylvania. The 20-cm thick mirror substrate was cast by Corning using ultra-low expansion (ULE) glass. The thermal expansion coefficient is 10 parts per billion per °C. The surface error is 12 nm after a 32-mode active correction of residual deformation over the entire mirror surface of 8.2 m in diameter; this is the most accurate mirror of such size ever produced. The focal length of the Subaru Telescope's primary mirror is 15 m so that its f-number is 1.83. The light gathering power of the Subaru Telescope is more than 10 times larger than that of the HST.

The conic constant of the primary mirror of the Subaru Telescope is −1.00835051. There are three secondary mirrors:

1. Cassegrain optical (Cass/Opt), which has a diameter of 1.330 m, focal length of −2.762 m, conic constant of −1.917322232, and located at a distance of 12.652174 m from the primary mirror
2. Nasmyth optical (Nas/Opt), which has a diameter of 1.400 m, focal length of −2.939 m, conic constant of −1.865055214, and located at a distance of 12.484300 m from the primary mirror
3. Cassegrain Infrared (Cass/IR), which has a diameter of 1.265 m, focal length of −2.762 m, conic constant of −1.917322232, and located at a distance of 12.652174 m from the primary mirror

2.6   ELLIPSOIDAL MIRRORS

The equation of an ellipse is given by

$$\frac{x^2}{a^2} + \frac{y^2}{b^2} = 1 \tag{2.13}$$

where a and b are the semi-major and semi-minor axes of the ellipse. The focal lengths f_1 and f_2 of the ellipsoidal mirror are given by

$$f_1 = a - \sqrt{a^2 - b^2} \tag{2.14}$$

$$f_2 = a + \sqrt{a^2 - b^2} \tag{2.15}$$

The first focal point is closer to the mirror than the second focal point. Therefore, the first focal length is smaller than the second focal length. Figure 2.19 shows the focal points of an ellipsoidal mirror. An object source placed at the first focal point will be imaged at the second focal point and vice versa.

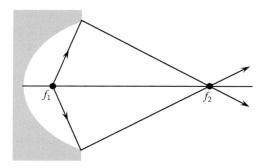

FIGURE 2.19   Two focal points of an ellipsoidal mirror.

The conic constant of an ellipse is given by

$$K = -1 + \frac{b^2}{a^2} \tag{2.16}$$

For given values of K and f_1 or f_2, one can determine the values of a and b. Note that for a spherical mirror, $K = 0, f_1 = f_2 = a$.

2.7   MIRROR ABERRATIONS

Mirrors are completely free of chromatic aberrations. A concave spherical mirror shows spherical aberration, coma, and astigmatism. A parabolic mirror is free of spherical aberration for an infinite distant axial object point. This is the reason parabolic mirrors are used in astronomical telescopes. However, a parabolic mirror shows significant astigmatism and coma for an off-axis object point.

2.8   HOMEWORK PROBLEMS

2.1 A plane parallel beam of light is incident upon a 30-mm diameter concave spherical mirror of 25 mm focal length ($f/0.83$); the incident light propagates along the axis of the spherical mirror. Calculate the focal length for rays, which are incident upon the mirror at a height of 15 mm from the axis of the mirror. Also, compute the percent change in the focal length of these rays compared to that for the paraxial rays.

2.2 A plane parallel beam of light is incident upon a 30-mm diameter parabolic mirror of 25 mm focal length ($f/0.83$); the incident light propagates along the axis of the parabolic mirror. Calculate the focal length for rays, which are incident upon the mirror at a height of 15 mm from the axis of the mirror, using slope of the tangent to the parabola at the diameter of the mirror.

2.3 A 3.0-mm long object is located at a distance of 40 mm from a concave spherical mirror of 25 mm focal length. Calculate the position and length of the image of the object.

2.4 The separation between the primary and secondary mirrors of the Hubble Space Telescope (HST) is 4.91 m. Calculate the diameter of the secondary mirror that is illuminated.

2.5 A plane parallel beam of light is incident upon a parabolic mirror at an angle of 5° with its axis. The diameter and focal length of the mirror are 30 mm and 25 mm, respectively. Calculate the focal length for rays, which are incident upon the mirror at a height of 15 mm from the axis of the mirror.

2.6 Show that the focal spot of the Hubble Space Telescope (HST) is located at a distance of 6.00 m from the secondary mirror.

2.7 The focal length of the Hubble Space Telescope (HST) is 57.6 m. Calculate the value of separation between the primary and secondary mirrors of the HST.

2.8 Calculate the values of a and b for the Subaru Telescope primary mirror and three secondary mirrors using the given values of the focal length and conic constants for these mirrors.

2.9 The focal length of the Subaru Telescope will depend upon the selection of a secondary mirror. Calculate the focal lengths of the Subaru Telescope for the three secondary mirrors.

2.10 Calculate the location of the focal spot as measured from the secondary mirror for each of the three secondary mirrors of the Subaru Telescope.

2.11 A small light source is placed at the second focal spot of an ellipsoidal mirror. If the image of this light source has a magnification of 1/6, then (1) show that the value of the conic constant is $-25/49$, and (2) determine the values of a and b when the value of f_1 is 25 mm.

3 Diffraction Gratings

3.1 INTRODUCTION

A diffraction grating is an optical component with a periodic structure, which produces periodic changes in the phase, amplitude, or both, of a monochromatic or polychromatic light beam incident upon it. David Rittenhouse, an American astronomer, produced the first diffraction grating around 1785 as a multiple-slit assembly consisting of fine wire. The multiple-slit assembly is a transmission amplitude grating. Diffraction gratings for optical spectroscopy are made by ruling grooves either on a transparent surface to obtain a transmission phase grating or metal surface to obtain a reflection phase grating. Holographic and dielectric gratings are produced using photolithography.

Light incident upon a grating is diffracted according to the grating equation

$$d\left(\sin\theta_m - \sin\theta_i\right) = m\lambda \tag{3.1}$$

where:

d is the grating groove spacing
λ is the wavelength of light
θ_i is the angle of incidence
θ_m is the angle of diffraction for order m that is equal to 0, ±1, ±2, and so on

Equation 3.1 may be written as

$$\theta_m = \sin^{-1}\left(\sin\theta_i + \frac{m\lambda}{d}\right) \tag{3.2}$$

To determine the number of diffraction orders recall that $-1 \le \sin\theta_m \le 1$. We obtain

$$-1 \le \sin\theta_i + \frac{m\lambda}{d} \le 1 \tag{3.3}$$

Hence

$$-(1+\sin\theta_i)\frac{d}{\lambda} \le m \le (1-\sin\theta_i)\frac{d}{\lambda} \tag{3.4}$$

The diffraction angle θ_0 for zeroth order is equal to the angle of incidence θ_i. The diffraction angle θ_{-1} is smaller than θ_i for the diffraction order $m = -1$ and the diffraction angle θ_1 is larger than θ_I for the diffraction order $m = 1$. This is illustrated for monochromatic light in Figures 3.1 and 3.2 for transmission and reflection gratings, respectively. It is interesting to note that if $d < \lambda$, then $-(1+\sin\theta_i) < m < 1-\sin\theta_i$. The only values of diffraction order m are equal to 0 and -1.

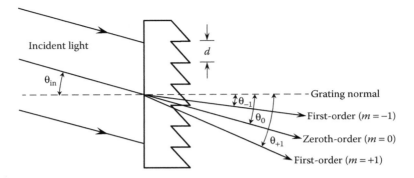

FIGURE 3.1 Diffraction angles for 0 and ±1 orders for a transmission grating.

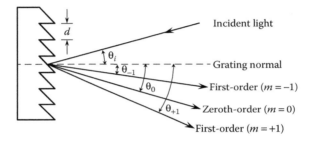

FIGURE 3.2 Diffraction angles for 0 and ±1 orders for a reflection grating.

The shape of the grooves is usually tailored to obtain most of the light intensity in the first order. Figure 3.3 shows the diffraction angles for wavelengths of 300, 500, and 700 nm for the $m = 1$ order for the light incident normally on a reflection grating with 1200 grooves/mm.

Some spectrometers use the Littrow configuration where θ_m is equal to $-\theta_i$, meaning the diffracted light is retro-reflected. The order m is negative for this configuration. The wavelength of the diffracted light is then given by

$$m\lambda = 2d \sin \theta_m \qquad (3.5)$$

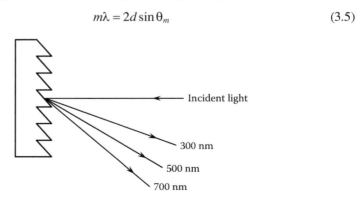

FIGURE 3.3 Diffraction angles for the $m = 1$ order of diffraction.

The resolving power $\lambda/\Delta\lambda$ of a grating in the Littrow configuration is obtained from Equation 3.5 as

$$\frac{\lambda}{\Delta\lambda} = \frac{\tan\theta_m}{\Delta\theta_m} \tag{3.6}$$

$$\Delta\theta_m = \frac{s_w}{f_s} \tag{3.7}$$

where s_w and f_s are the slit width and focal length of the spectrometer, respectively. Equations 3.6 and 3.7 yield a value of 2.5×10^4 for the resolving power of the spectrometer using values of 0.50 m for f_s, 10 μm for s_w, and 0.5 for $\tan\theta_m$. This value of 2.5×10^4 for the resolving power gives a value of 0.02 nm for $\Delta\lambda$ using a value of 500 nm for λ. Note that s_w must be greater than $\lambda f_s/D$, where D is equal to the footprint of the light beam on the grating. Therefore, the resolving power increases linearly with D. The maximum value of the resolving power of a grating is given by

$$\frac{\lambda}{\Delta\lambda} = Nm \tag{3.8}$$

where N is the number of the grating grooves that are illuminated by the incident light beam. Equation 3.8 yields a value of 6×10^4 for the resolving power of the grating using values of 6×10^4 for N, corresponding to 1200 grooves/mm with 50 mm wide beam of light incident upon the grating, and 1 for the diffraction order m.

Equation 3.8 may be expressed in the form

$$\frac{\lambda}{\Delta\lambda} = \frac{Nd\left(\sin\theta_m - \sin\theta_i\right)}{\lambda} \tag{3.9}$$

3.2   RULED GRATINGS

Master diffraction gratings are ruled with mechanically controlled ruling engines. Rowland is considered to be the father of modern ruled diffraction grating. In 1882, he ruled gratings of 6-inch widths with more than 100,000 grooves and resolving power in excess of 1.5×10^5 (Harrison 1949). Rowland and his successors at the Johns Hopkins University designed and implemented a series of ruling engines, each superior to its predecessor. Rowland engines were capable of producing gratings as large as 7.5 inches. Michelson was able to produce gratings as large as 10 inches. One of Michelson's engines later ended up in Prof. Harrison's laboratory at Massachusetts Institute of Technology in 1947. Harrison and his team equipped their engine with interferometric position feedback control (Harrison and Stroke 1955). Harrison's design has become standard practice in modern ruling engines. Contemporary gratings are ruled in thin films of aluminum evaporated on optically flat glass blanks. Because aluminum is fairly soft, it causes less wear on the ruling diamond tool.

Commonly used ruled diffraction gratings are the relatively inexpensive metalized replicas of the master diffraction gratings. Replica gratings are made by pouring a plastic solution over an original grating, evaporating the solvent, and removing the resulting film, which has the grooves of the original grating impressed upon it (Wallace 1905). The removed film is then attached to a substrate. The groove density in the replica is somewhat higher than that in the original grating because of shrinkage of the replica.

Grooves of a diffraction grating are usually shaped to obtain most of the diffracted energy in a given order m for the desired wavelength λ. Such gratings are referred to as blazed reflection gratings. Figure 3.4 shows a blazed reflection grating with triangular groove facets, which are at an angle γ (called the blaze angle) with the grating surface.

For $\theta_i = \gamma$, most of the diffracted energy goes in the direction of the specular reflection from the groove facets, which corresponds to $\theta_m = -\gamma$. In this case, the grating equation becomes

$$2d \sin \gamma = m\lambda \tag{3.10}$$

For $m = 1$, the blaze wavelength is given by

$$\lambda_B = 2d \sin \gamma \tag{3.11}$$

For many commercial gratings, the value of $\sin \gamma$ is equal to 0.45, corresponding to $\gamma = 26°45'$, so that λ_B is equal to $0.9d$. For example, if there are 1800 grooves/mm, the value of d is $5/9$ µm. For this value of d, the value of λ_B is 0.5 µm or 500 nm.

Diffraction efficiency η of a grating depends upon the wavelength of light and its polarization, relative to the grooves of the grating. Figure 3.5 shows typical diffraction efficiency of a 500-nm blazed ruled diffraction grating with 1800 grooves/mm.

Echelle gratings are used for high-resolution spectroscopy. They are coarsely ruled (large d) blazed gratings with a large blaze angle. The resolving power of an echelle grating used in the Littrow mode at the blaze angle is proportional to the tangent of the large blaze angle. Echelle gratings are generally used with a second grating, or prism, to separate the overlapping diffraction orders. Albert Michelson, who referred to them as echelons, discovered echelle gratings in 1898. The resolving

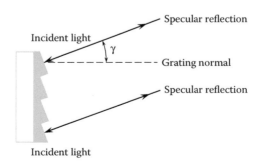

FIGURE 3.4   Blazed grating.

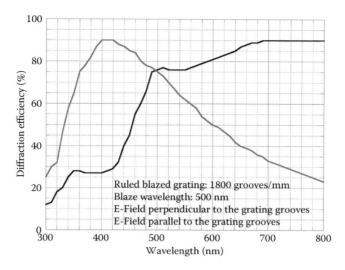

FIGURE 3.5 Efficiency of an 1800-grooves/mm ruled grating. (Courtesy of Newport Corporation.)

power of an echelle grating relative to that of a conventional blazed grating in the Littrow mode is given by

$$\frac{(\lambda/\Delta\lambda)_E}{(\lambda/\Delta\lambda)_C} = \frac{\tan\theta_E}{\tan\theta_C} \tag{3.12}$$

where θ_E and θ_C are the angles of diffraction for the echelle and conventional gratings, respectively. For values of 75° and 30° for θ_E and θ_C, the resolving power of an echelle grating relative to that a conventional grating is equal to 6.5. In other words, the resolving power of the echelle grating with θ_E of 75° is 6.5 times larger than that of the conventional grating with θ_C of 30°. That is why echelle gratings are used for high-resolution spectroscopy. Figure 3.6 shows an echelle grating used in the Littrow mode.

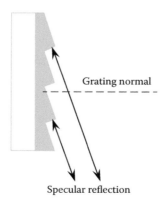

FIGURE 3.6 Echelle grating in the Littrow mode.

3.3   HOLOGRAPHIC GRATINGS

Holographic gratings are produced using interference photolithography, which eliminates the periodic errors found in ruled gratings. Horiba Jobin Yvon produced the first holographic grating in 1967 using two collimated laser beams of the same wavelength that are incident at angle θ upon a photosensitive layer deposited on a highly-polished substrate, as shown in Figure 3.7 (Lerner et al. 1980). A team of Massachusetts Institute of Technology (MIT) students, scientists, and engineers have developed the world's fastest and most precise tool, the Nanoruler, for producing large holographic gratings (Chen 2004). The Nanoruler can fabricate holographic gratings 10–1000 times faster and more precisely than previous methods.

Holographic gratings have two groove profiles: sinusoidal and blazed. Most of the holographic gratings have sinusoidal groove profiles, which are symmetrical and therefore have no blaze direction. The efficiency of a holographic grating with sinusoidal groove profile is usually less than that of a ruled grating. However, a holographic grating with a sinusoidal groove profile has nearly the same efficiency as a blazed ruled grating when the groove spacing is comparable to the wavelength of light. A blazed holographic grating has a saw-tooth groove profile, which has higher efficiency than that of a holographic grating with sinusoidal groove profile. Replicas of holographic master gratings are produced using a process identical to that used for ruled gratings. Figure 3.7 shows a schematic for the recording of a holographic grating.

The spatial period of the grating is $d = \lambda/(2\sin\theta)$. Hence for $\theta > \sin^{-1}(1/2)$, the grating period becomes less than λ. In the extreme case, grazing angle $d = \lambda/2$.

Figure 3.8 shows the efficiency of a holographic grating with 1200 grooves/mm. Note that the efficiency of the holographic grating shown in Figure 3.8 is quite comparable to that of the ruled grating shown in Figure 3.5.

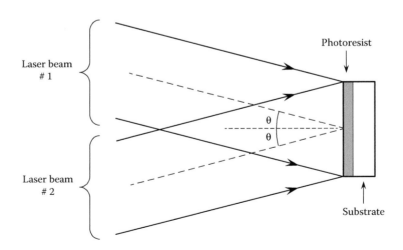

FIGURE 3.7   Recording of a holographic grating.

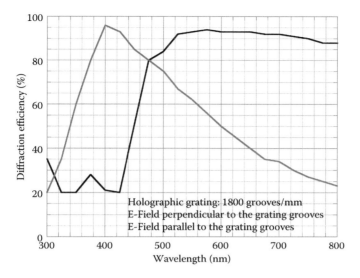

FIGURE 3.8 Efficiency of an 1800-grooves/mm holographic grating. (Courtesy of Newport Corporation.)

3.4 MULTILAYER DIELECTRIC GRATINGS

Multilayer dielectric gratings provide high diffraction efficiency (>96%) for TE-polarized light in the $m = -1$ order in reflection and high damage threshold above that achievable with metallic gratings. The diffraction efficiency is low (<50%) for TM-polarized light (Perry et al. 1995). Figure 3.9 shows the basic multilayer dielectric grating structure, which consists of alternate layers of materials of high (dark bands) and low (white bands) refractive indexes. The groove height and the top layer thickness are denoted by h and t, respectively. High diffraction efficiency occurs when the optical path of the grooves is near one quarter of a wave and the optical thickness of the top layer is near three quarters of a wave. The peak diffraction

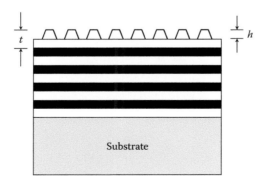

FIGURE 3.9 Structure of a multilayer dielectric grating.

efficiency (order $m = 1$) has been predicted to be near 98% for TE-polarized light and less than 50% for TM-polarized light at 1053 nm of a grating consisting of alternate layers of ZnS ($n = 2.35$) and ThF$_4$ ($n = 1.52$). Multilayer dielectric gratings could be designed for any wavelength using appropriate materials.

3.5 HOMEWORK PROBLEMS

3.1 A collimated beam of 514.5 nm Hg green light incident normally upon a grating is diffracted through an angle of 55.4° for the 2nd order. Calculate the number of grating grooves/mm.

3.2 A 532-nm green laser beam is incident at an angle of 20° upon a grating with 500 grooves/mm. Determine the highest diffraction order that can be observed.

3.3 A Hg spectrum contains a blue line at 435.8 nm and a green line at 546.1 nm. A 1200-grooves/mm grating diffracts these lines in the first order. The angle of incidence with the grating is 15°. Determine the angular separation between the 435.8- and 546.1-nm diffracted beams.

3.4 Collimated light of 550 and 560 nm wavelengths is incident normally upon a plane ruled grating with 500 grooves/mm. A 1.0 m focal length is used to focus the diffracted light in the 1st order. Determine the separation between the 550 and 560 nm focal spots in the focal plane of the lens.

3.5 A 5-cm wide plane diffraction grating with 1200 grooves is used in the 2nd order for 500-nm light. Determine (1) resolving power and (2) spectral resolution of the grating.

3.6 A plane diffraction grating is used to resolve two spectral components in the 3rd order near 500 nm. Determine the minimum number of grating grooves required to resolve the two components, which are separated by 0.01 nm.

3.7 Determine the diffraction efficiency of the holographic grating at 600 nm for TE- and TM-polarizations, using the efficiency curves in Figure 3.8.

3.8 A 630 nm light diffracted by a plane grating in a given order overlaps a 420 nm light diffracted by the same grating in the next higher order. Determine the values of the two-diffraction orders.

3.9 A high-resolution grating with 1800 grooves/mm has resolving power of 1×10^6 for 500 nm in the 3rd order. Determine the width of the grating.

4 Polarizers

4.1 INTRODUCTION

Light is an electromagnetic wave. The electric field **E** of this wave oscillates perpendicular to the direction of propagation. At a given point in space, polarization of light is defined as the shape of the curve traced out with time by the tip of **E**. For example, if the tip of **E** must move along a straight line with time, such light is linearly polarized. Similarly, if the tip of **E** traces a circle with time, we have circularly polarized light. In general, light may be elliptically polarized with linear and circular polarizations being the limiting cases of the elliptic polarization. If the polarization of light exhibits random fluctuations in time so that its temporal average is zero over a long time interval (>> 1/frequency of light), such light is said to be unpolarized. A linear polarizer produces linearly polarized light using unpolarized light. The transmittance T of an ideal polarizer for unpolarized light is 50%. The transmittance of a real polarizer is < 50% for unpolarized light.

Malus' law gives transmittance of an ideal linear polarizer for linearly polarized light as

$$T = (\cos\theta)^2 \qquad (4.1)$$

where θ is the angle of the transmission axis of the linear polarizer with the axis of the linearly-polarized light. T would be equal to 1 for $\theta = 0°$ and 0 for $\theta = 90°$.

Extinction ratio ER is the ratio of transmittance of the desired polarization to the transmittance of the undesired polarization. ER is given by

$$\mathrm{ER} = \frac{T_{\max}}{T_{\min}} \qquad (4.2)$$

where $T_{\max}$ and $T_{\min}$ are the maximum and minimum values of T. Values of ER can range from 10^2 for sheet polarizers to 10^6 for birefringent polarizers. Other important characteristics of a polarizer are its acceptance angle and clear aperture. The acceptance angle is the largest deviation from the design incidence angle at which the polarizer will perform within specifications. Clear aperture is most restrictive for birefringent polarizers. Dichroic polarizers have the largest clear apertures.

Linear polarizers may be classified into three categories: birefringent, dichroic, and reflective. Birefringent polarizers separate the two polarization states of the incident light using a birefringent crystalline material. Dichroic polarizers are absorptive polarizers, which transmit one polarization state and absorb the orthogonal polarization state. Reflective polarizers reflect one polarization state and transmit the orthogonal polarization state.

4.2 BIREFRINGENT LINEAR POLARIZERS

Birefringent linear polarizers are made of uniaxial crystals, which belong to hexagonal, tetragonal, and trigonal crystal systems. Uniaxial crystals have two indices of refraction. One of these indices of refraction is called the ordinary index of refraction n_O, which stands for polarization perpendicular to the uniaxial crystal axis called the optic axis. The ordinary (O) wave obeys Snell's law of refraction. The other index of refraction, the extraordinary index of refraction n_E, represents polarization parallel to the optic axis. The effective index of refraction n_e for the extraordinary wave will lie between the values of n_O and n_E, which is determined by the direction of propagation in the crystal. The magnitude of birefringence is given by

$$\Delta n = n_E - n_O \qquad (4.3)$$

Birefringent crystal is a uniaxial positive crystal if Δn is positive. The birefringent crystal is a uniaxial negative crystal if Δn is negative. Calcite is a widely used material for birefringent polarizers. The values of n_O and n_E for calcite are 1.6584 and 1.4864 at 590 nm, respectively. Therefore, calcite is a uniaxial negative crystal.

We consider the following calcite birefringent linear polarizers: (1) Nicol prism, (2) Glan-Thompson, (3) Glan-Taylor, (4) Wollaston prism, (5) Rochon prism, and (6) Senarmont prism. The transmittance of these polarizers is > 85% for the E ray. Glan-Taylor polarizers have a higher optical damage threshold than Glan-Thompson due to an air gap instead of cement in between the two constituent prisms.

4.2.1 NICOL CALCITE PRISM LINEAR POLARIZER

An early example of a birefringent polarizer is the Nicol calcite prism linear polarizer, which consists of a crystal of calcite that has been split and rejoined with Canada balsam with refractive index of 1.55 at 590 nm. Figure 4.1 shows a schematic of the Nicol calcite prism linear polarizer (Jenkins and White 1976).

Angles of the calcite prisms are 68°, 90°, and 22°. The optic axis lies in the plane of the schematic at an angle of 48° with the AB edge. For unpolarized light incident upon the first calcite prism along the AC/BD edges, the E ray propagates in the first calcite prism at an angle of 56.1° with the optic axis. The refractive index n_e for the

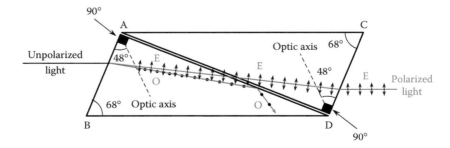

FIGURE 4.1   Nicol calcite prism linear polarizer.

E ray is 1.534, which is close to but less than that of the Canada balsam. The E ray partly refracts into the balsam and on through the second calcite prism into air. The critical angle for total internal reflection of the O ray at the first calcite prism to balsam interface is about 69°, which is smaller than the 77° angle of incidence of the O ray at the first calcite-balsam interface. Hence, the O ray is totally reflected into the first calcite prism. The critical angle for total internal reflection at the calcite-air interface is 37°, which is less than the 55° angle of incidence of the O ray at the calcite-air interface. Therefore, the O ray undergoes total internal reflection at the calcite-air interface. As a result, the O ray is unable to escape the Nicol prism.

4.2.2 GLAN-THOMPSON CALCITE PRISM LINEAR POLARIZER

Glan-Thompson calcite linear polarizers are made with end faces cut perpendicular to the sides so that the light enters and leaves normal to the end faces. These polarizers cover the wavelength range 350–2200 nm. The extinction ratio is $> 2 \times 10^5$. Similar to the Nicol prism, Figure 4.2 shows a schematic of a Glan-Thompson calcite linear polarizer. It consists of two right-angled calcite prisms, which are cemented together with Canada balsam or a synthetic polymer. The optic axes of the two prisms are parallel but perpendicular to the plane of the schematic. The refractive index of the cement (1.55) is higher than that of calcite for the E ray (1.4864) but lower than that for the O ray (1.6584). Therefore, the E ray is transmitted from the first prism into cement and then out into the air through the second prism without suffering any displacement in its position. On the other hand, the O ray undergoes total internal reflection at the calcite-cement interface because its angle of incidence is equal to or greater than the critical angle (69°) for total internal reflection. The length to width ratio L/W of the Glan-Thompson prism must be > 2.6 in order to have total internal reflection for the O ray assuming refractive index of the cement to be 1.55. The acceptance angle is 15°–30° (Edmund Optics 2017).

4.2.3 GLAN-TAYLOR CALCITE PRISM LINEAR POLARIZER

Glan-Taylor calcite prism linear polarizers cover the same wavelength range as that of the Glan-Thompson calcite prism linear polarizers. However, the Glan-Taylor calcite prism linear polarizers have higher damage threshold than that of the Glan-Thompson calcite prism linear polarizers because its two right-angled calcite prisms

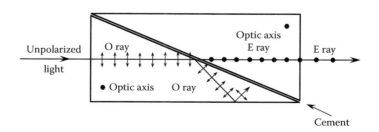

FIGURE 4.2 Glan-Thompson calcite linear polarizer.

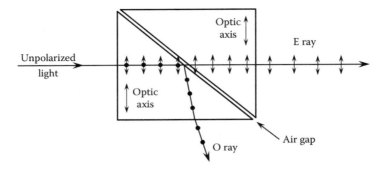

FIGURE 4.3 Glan-Taylor calcite prism linear polarizer.

are not cemented but have an air gap between them. Figure 4.3 shows a schematic of a Glan-Taylor calcite prism linear polarizer ($L/W = 1.2$). The optic axes of the two right-angled calcite prisms are parallel to the input face. The critical angle for total internal reflection at the calcite-air interface for the E and O rays are 42° and 37°, respectively. The Glan-Taylor calcite prism linear polarizer must satisfy the condition $1.11 < L/W < 1.33$ in order to obtain transmission of the E ray and total internal reflection of the O ray at the calcite-air interface. The acceptance angle is 6° (Edmund Optics 2017).

4.2.4 WOLLASTON CALCITE PRISM LINEAR POLARIZER

Wollaston calcite prism polarizers separate unpolarized light into two orthogonal linearly polarized beams that leave the polarizer through the same exit port. Commercial polarizers are available with divergence angles δ from about 15° to 45° between the two beams. The Wollaston calcite prism linear polarizer consists of two right-angled calcite prisms, which are cemented together as the Glan-Thompson calcite prism polarizer. However, the optic axes of the two prisms are orthogonal to each other. Figure 4.4 shows a Wollaston calcite prism linear polarizer with L/W equal to 1.0.

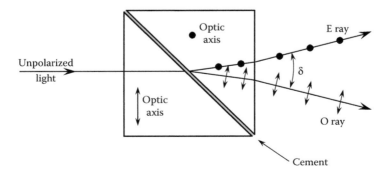

FIGURE 4.4 Wollaston calcite prism linear polarizer.

The optic axis of the first prism is parallel to the width in the plane of the schematic. The optic axis of the second prism is perpendicular to the plane of the schematic. The E ray in the first prism becomes the O ray in the second prism; this ray enters a medium of high refractive index (1.6584) from a medium of low refractive index (1.4864). Similarly, the O ray in the first prism becomes the E ray in the second prism; this ray enters a medium of low refractive index (1.4864) from a medium of high refractive index (1.6584). This implies that the two rays, which propagate together in the first prism, are split in opposite directions in the second prism. The divergence angle between the two rays in the second prism is about 13°. The divergence angle between the two rays further increases to about 20° upon exiting into air from the second prism. The acceptance angle is 20° (Edmund Optics 2017).

4.2.5 ROCHON CALCITE PRISM LINEAR POLARIZER

The Rochon calcite prism linear polarizer is similar to the Wollaston calcite prism linear polarizer except that the optic axis in the first calcite prism is parallel to the length of the polarizer, as shown in Figure 4.5 for L/W equal to 1. Because of the orientation of the optic axes in the calcite prisms, the O ray has the same refractive in both the prisms. Therefore, the O ray propagates through the Rochon polarizer without any deviation; however, the E ray suffers the same divergence as in the Wollaston polarizer. The acceptance angle is 20° (Edmund Optics 2017).

4.2.6 SENARMONT CALCITE PRISM LINEAR POLARIZER

The Senarmont calcite prism linear polarizer is similar to the Wollaston calcite prism linear polarizer except that the optic axes in the Senarmont prism are along the length and width of the polarizer, as shown in Figure 4.6.

The O ray has the same refractive index in both the prisms and propagates through the Senarmont polarizer without any deviation. The E ray suffers the same deviation as in the Wollaston and Rochon polarizers. However, in the Senarmont polarizer, the E-fields of the O and E rays are orthogonal to those in the Wollaston and Rochon polarizers.

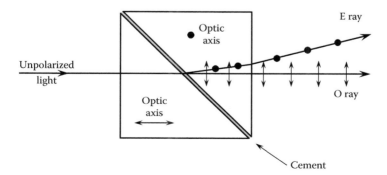

FIGURE 4.5   Rochon calcite prism linear polarizer.

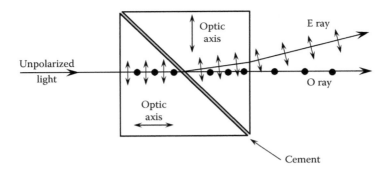

FIGURE 4.6   Senarmont calcite prism linear polarizer.

4.3   DICHROIC LINEAR POLARIZERS

There are a number of dichroic crystalline materials such as tourmaline and herapa-thite (iodosulfate of quinine), which strongly absorb light of one polarization and transmit light of orthogonal polarization. For that reason, dichroic materials can be used for producing linear polarizers. Tourmaline is not used as a polarizer because the dichroic effect is strongly wavelength dependent. It is difficult to grow large crystals of herapathite.

Polaroid is the trade name for the most commonly used dichroic material, which is a plastic sheet of iodine-impregnated polyvinyl alcohol (PVA). When a PVA sheet is heated and stretched in one direction, the long polymeric molecules align in the direction of the stretch. The iodine atoms provide electrons, which move easily along the aligned chains but not perpendicular to them. Therefore, light with **E**-field paral-lel to the aligned chains is heavily absorbed, and the light with **E**-field perpendicular to the aligned chains is transmitted. Land includes a summary of the development of sheet polarizers (Land 1951).

4.3.1   HN POLAROID SHEET LINEAR POLARIZERS

There are several types of HN Polaroid sheet linear polarizers: HN42, HN38, HN32, and HN22, which cover the visible range 400–700 nm. HP42 has been developed as a direct replacement for HN22, HN32, HN38, and HN42 and is commercially avail-able (VisionTech Systems 2017). The transmittance of HP42 for unpolarized light is 35%–45% in the wavelength range 450–800 nm.

4.3.2   HR POLAROID SHEET LINEAR POLARIZERS

HR Polaroid sheet linear polarizers cover the near-infrared (NIR) wavelength range 800–2700 nm. The standard HR polarizer is a plastic lamination.

4.3.3 POLARCOR GLASS LINEAR POLARIZER

Polarcor, developed by Corning, consists of elongated silver crystals in borosilicate glass. The polarization is produced by the absorption of the aligned elongated silver crystals. It has high transmittance and high extinction ratio in the wavelength range 600–2300 nm.

4.4 REFLECTIVE LINEAR POLARIZERS

Reflective linear polarizers such as Brewster windows and wire-grid polarizers transmit the desired polarization and reflect the orthogonal polarization.

4.4.1 BREWSTER WINDOW POLARIZER

Brewster angle θ_B for a dielectric window of refractive index n is given by

$$\theta_B = \tan^{-1} n \tag{4.4}$$

When unpolarized light is incident upon an uncoated surface of a dielectric window at the Brewster's angle, as shown in Figure 4.7, the reflectance for light with **E**-field in the plane of incidence (TM polarization) is zero and the reflectance for light with **E**-field perpendicular to the plane of incidence (TE polarization) is finite.

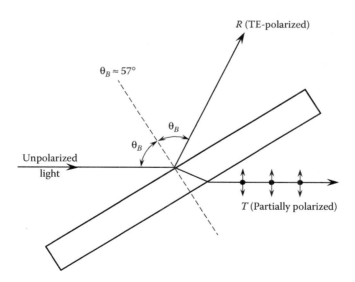

FIGURE 4.7   Brewster-angle glass window linear polarizer.

Hence, the reflected light is completely polarized (TE polarization) while the transmitted light is only partially polarized. The reflectance R for the TE polarization is given by

$$R_{TE} = \left(\frac{n^2-1}{n^2+1}\right)^2 \tag{4.5}$$

For a dielectric material of refractive index 1.5 such as glass, R_{TE} is equal to 14.8%.

A single Brewster window will produce a 100% TE-polarized reflected beam, but a single Brewster window is not sufficient for the transmitted TM-polarization. However, a single Brewster window placed in a laser cavity can produce a TM-polarized light because of the many passes through the Brewster window in the laser cavity.

4.4.2 BREWSTER PILE-OF-PLATES LINEAR POLARIZER

The degree of polarization of the transmitted beam can be increased using a number of Brewster plates in tandem. This arrangement is called a Brewster pile-of-plates linear polarizer. Figure 4.8 shows a Brewster pile-of-plates linear polarizer consisting of three plates. The degree of polarization of the transmitted light is given by (Jenkins and White 1976)

$$P = \frac{m}{m + \left(2n / n^2 - 1\right)^2} \tag{4.6}$$

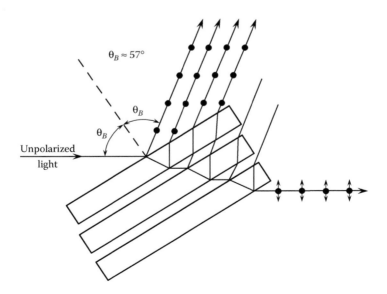

FIGURE 4.8 Brewster pile-of-plates linear polarizer.

where m is the number of plates. Equation 4.6 also takes into account rays internally reflected two or more times. The value of P is 90% for 50 plates with refractive index of 1.5.

4.4.3 WIRE-GRID LINEAR POLARIZER

A wire-grid linear polarizer consists of many parallel thin metal wires on a substrate. Light polarized along the direction of these wires (TE polarization) is reflected while light perpendicular to the direction of these wires (TM polarization) is transmitted. Wire-grid polarizers cover a very large wavelength range well into the infrared (IR). The wavelength range is limited by the transmittance of the substrate. Numerous wire-grid linear polarizers are commercially available for use in the UV, VIS, and IR regions. An aluminum-wire grid layered between two glass windows is useful for the wavelength range 400–700 nm. Aluminum-wire grid layered between two fused silica windows is useful for the wavelength range 300–3000 nm. Gold-wire 2880 lines/mm grid, which is vapor deposited on a silver bromide substrate, is useful for the IR wavelength range 2.5–35 μm (Perkin-Elmer 2017). CaF_2, BaF_2, ZnSe, KRS-5, and Ge substrates are used for the IR wavelength range 2–8 μm, 2–12 μm, 2–19 μm, 2–30 μm, and 8–17 μm, respectively (Edmund Optics 2017). The extinction ratio of these IR wire-grid polarizers is in the range 50–300.

4.5 CIRCULAR POLARIZERS

Light can be circularly polarized using a phase difference of 90° between the orthogonal polarization components of light, such as in a Fresnel rhomb or a quarter-wave plate.

4.5.1 FRESNEL RHOMB CIRCULAR POLARIZER

A Fresnel rhomb is made of a material with refractive index n. The phase changes that occur in the p and s polarization components upon total internal reflection are given by

$$\tan \frac{\delta_p}{2} = \frac{n\sqrt{n^2 \sin^2 \theta_i - 1}}{\cos \theta_i} \tag{4.7}$$

$$\tan \frac{\delta_s}{2} = \frac{\sqrt{n^2 \sin^2 \theta_i - 1}}{n\cos \theta_i} \tag{4.8}$$

where:

δ_p and δ_s are the phase changes for the p and s polarizations, respectively
θ_i is the angle of incidence

Equations 4.7 and 4.8 are used to determine θ_i so that the phase difference δ between the p and s polarizations is equal to 45°. Because there are two internal reflections in a Fresnel rhomb, the total value of δ is 90° as required.

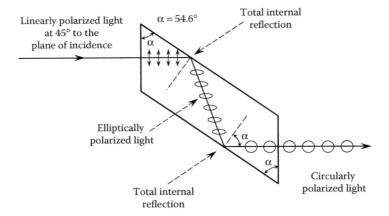

FIGURE 4.9   Fresnel glass rhomb circular polarizer.

Figure 4.9 shows a Fresnel rhomb of glass, which has a refractive index of 1.51. The value of δ is 45° for θ equal to 54.6° (Jenkins and White 1976).

Linearly polarized light with **E**-field at 45° to the plane of incidence is incident normally on the shorter surface of the rhomb. Then it is incident upon the first larger surface of the rhomb at an angle incidence of 54.6°, where it is totally reflected with a phase difference of 45° between the p and s polarization components leading to elliptically polarized light. Another total internal reflection at the second large surface of the rhomb produces an additional phase difference of 45° between the p and s polarization components and results in circularly polarized light. Light is right circularly polarized if the **E**-field of the incident light is at an angle of 45° with the direction of p polarization as shown in Figure 4.10. The Fresnel rhomb has the disadvantage that the output beam is displaced from the input beam.

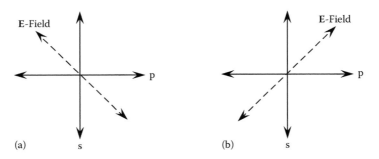

FIGURE 4.10   Orientation of E-field of the incident light for right (a) and left (b) circular polarizations.

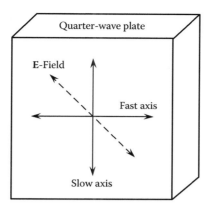

FIGURE 4.11   Quarter-wave plate right circular polarizer.

4.5.2   QUARTER-WAVE PLATE CIRCULAR POLARIZER

A quarter-wave plate (QWP) of a birefringent crystal such as calcite is used to provide the required phase shift of 90° between the two orthogonal components of linearly polarized light. In propagating through a combination of the linear polarizer and QWP, unpolarized light produces right circularly polarized light, if the transmission axis of the linear polarizer is set +45° with the fast axis of the QWP as shown in Figure 4.11. Left circularly polarized light is obtained if the transmission axis of the linear polarizer is set at −45° with the fast axis of the QWP. Polaroid developed HNCP37 sheet circular polarizers, once available in 12″ square sheets, are no longer available. But smaller sheets are available from other vendors.

A half-wave plate (HWP) has twice the thickness of a QWP. A HWP is used to rotate the linear polarization through a desired angle.

4.6   JONES MATRIX ALGEBRA

Jones matrix algebra is used to determine the effect of a given optical component upon the polarization of light transmitted through it. A 1 × 2 matrix, called a Jones vector, is used to specify the polarization of light. A 2 × 2 matrix, called a Jones matrix, is used to represent an optical component.

The Jones vector for linearly polarized (LP) light at angle θ with the x-axis is given by (Fowles 1975)

$$J_{LP} = \begin{bmatrix} A \\ B \end{bmatrix} \tag{4.9}$$

where θ is equal to $\tan^{-1}(B/A)$.

The Jones vector for right circularly polarized (RCP) light is given by

$$J_{RCP} = \begin{bmatrix} 1 \\ i \end{bmatrix} \tag{4.10}$$

The Jones vector for left circularly polarized (LCP) light is given by

$$J_{LCP} = \begin{bmatrix} 1 \\ -i \end{bmatrix}$$ (4.11)

The Jones matrix M for an optical component is defined by

$$J_2 = MJ_1$$ (4.12)

where:
J_1 is the Jones vector for the light incident upon the optical component
J_2 is the Jones vector for the light transmitted through the optical component

The Jones matrix for an ideal linear polarizer at an angle θ with the x-axis is given by

$$M_{LP} = \begin{bmatrix} \cos^2\theta & \cos\theta\sin\theta \\ \cos\theta\sin\theta & \sin^2\theta \end{bmatrix}$$ (4.13)

The Jones matrix for an ideal right circular polarizer is given by

$$M_{RCP} = \frac{1}{2}\begin{bmatrix} 1 & -i \\ i & 1 \end{bmatrix}$$ (4.14)

The Jones matrix for an ideal left circular polarizer is given by

$$M_{LCP} = \frac{1}{2}\begin{bmatrix} 1 & i \\ -i & 1 \end{bmatrix}$$ (4.15)

The Jones matrix for a quarter-wave plate with the fast axis at an angle θ with the x-axis is given by

$$M_{QWP} = \begin{bmatrix} \cos^2\theta + i\sin^2\theta & (1-i)\sin\theta\cos\theta \\ (1-i)\sin\theta\cos\theta & \sin^2\theta + i\cos^2\theta \end{bmatrix}$$ (4.16)

If light passes through a system of optical components with Jones matrices M_1, M_2, M_3..., the Jones matrix for the system is given by

$$M = ...M_3M_2M_1$$ (4.17)

4.7 HOMEWORK PROBLEMS

4.1 Unpolarized light is incident upon a linear polarizer with maximum transmittance of 90% for linearly polarized light. A second polarizer, identical to the first polarizer, is placed after the first polarizer. The angle between the transmission axes of the two polarizers is θ. Determine the fraction of the unpolarized light that is transmitted through this combination of the two polarizers for θ equal to 30°, 45°, and 60°.

4.2 Linearly polarized light is incident upon a linear polarizer, which is rotated about its normal axis at 50 cycles per second. Determine the frequency at which intensity of the light transmitted through the linear polarizer is modulated.

4.3 Show that when light is incident upon a plane-parallel plate of a material at the Brewster angle, the refracted beam is also incident on the second surface at its Brewster angle.

4.4 A Brewster pile-of-plates polarizer for use in the infrared consists of 5 germanium plates of refractive index 4.0. Determine the degree of polarization of the light transmitted by this polarizer.

4.5 The critical angle for total internal reflection in a material is 45°. Determine the value of the Brewster angle for this material.

4.6 A Fresnel rhomb for use in the infrared is made of CsI, which has a refractive index of 1.70 at 32 μm. Determine the angle of incidence θ_i at which the phase difference δ between the p and s polarization components is 45°.

4.7 A calcite quarter-wave plate is designed for use at 1.06 μm. Determine its physical thickness, assuming refractive indices n_o and n_E are equal to 1.6584 and 1.4864.

4.8 A half-wave plate consists of two quarter-wave plates. Determine the Jones matrix for the half-wave plate with the fast axis at an angle θ with the x-axis.

4.9 A right circular polarizer includes a linear polarizer at 45° with the x-axis and a quarter-wave plate at 0° with the x-axis. Determine the Jones matrix for the combination of the linear polarizer and the quarter-wave plate.

4.10 Show that sending unpolarized light through a linear polarizer and a quarter-wave plate only in the right order produces circularly polarized light.

4.11 Linearly polarized light along the x-axis is sent through two linear polarizers. The first is oriented with its transmission axis at 45° with the x-axis and the second with its transmission axis vertical (perpendicular to x-axis). Show that the transmitted light is linearly polarized in the vertical direction.

5 Optical Windows

5.1 INTRODUCTION

Many optical systems require windows to protect the system from dust and debris. Low-temperature cryostats and ultra-high vacuum (UHV) thin film deposition systems, such as molecular beam epitaxy (MBE), require windows to keep the environment in the system separated from the surrounding environment. In the latter cases, atmospheric pressure is exerted on the outside surface of the window while the inside surface is under vacuum. As a result, the window in the cryostat, or the MBE system, must withstand the pressure difference between the atmosphere from the outside and the vacuum on the inside. The design of a window involves several factors such as strength, transmission range, environmental durability, and availability of anti-reflections coatings. MBE windows are made of glass sealed onto stainless steel conflate flanges with a vacuum integrity of $5–10 \times 10^{-11}$ torr.

The minimum thickness t_w of an unclamped window of radius R_w under pressure differential ΔP_w is given by (Harris 1999)

$$t_w = \left(\frac{1.25 \Delta P_w S_F}{M_R} \right)^{1/2} R_w \tag{5.1}$$

where:

M_R is modulus of rupture
S_F is the safety factor

Table 5.1 lists values of M_R for several window materials.

Taking into consideration a fused silica window with $\Delta P_w = 0.10$ MPa and $R_w = 25$ mm and using a value of 4 for S_F, then Equation 5.1 yields a value of 2.1 mm for t_w. Values of the modulus of rupture in Table 5.1 should be considered approximate because they vary with the quality of surface finish, method of fabrication, material purity, and size of the window.

Neglecting interference effects, external transmittance T_E of a plane parallel window for normal incidence of light is given by

$$T_E = \left[\frac{(1-R)^2}{1 - R^2 \, e^{-2\alpha t_w}} \right] e^{-\alpha t_w} \tag{5.2}$$

where:

R is the single-surface reflectance
α is the bulk absorption coefficient of the window material

TABLE 5.1
M_R **for Window Materials**

Material	BaF_2	CaF_2	FS	Ge	Al_2O_3	Si	ZnSe
M_R (MPa)	27	37	60	90	300	120	50

Equation 5.2 is valid for $\alpha t_w \cdot 1$. Windows are antireflection (AR) coated for increased external transmittance. Assuming perfect AR coating, external transmittance is equal to the internal transmittance T_I, which is given by

$$T_I = e^{-\alpha t_w} \tag{5.3}$$

It is important to have a low value of αt_w in order to obtain a high value of the internal transmittance of the window. A number of materials are commercially available that have low values of α for UV (200–400 nm), VIS (400–700 nm), NIR (700–1500 nm), and IR (1,500–50,000 nm) windows. Some of these materials are listed in Table 5.2.

TABLE 5.2
Commercially Available Window Materials

Material	Spectral Range	Refractive Index
UV Grade fused silica (SiO_2)	UV, VIS, and NIR	1.462 at 500 nm
IR Grade fused silica (SiO_2)	VIS, NIR, and IR	1.462 at 500 nm
Barium fluoride (BaF_2)	UV, VIS, and IR	1.478 at 500 nm
Calcium fluoride (CaF_2)	UV, VIS, and IR	1.437 at 500 nm
Lithium fluoride (LiF)	UV, VIS, and IR	1.394 at 500 nm
Magnesium fluoride (MgF_2)	UV, VIS, and IR	1.380 at 500 nm
N-BK7 Glass	VIS and NIR	
Lithium niobate ($LiNbO_3$)	VIS and IR	2.286 (n_O) at 633 nm
		2.203 (n_E) at 633 nm
Magnesium oxide (MgO)	VIS and IR	1.74 at 500 nm
Sapphire (Al_2O_3)	VIS and IR	1.755 (n_O) at 1.0 μm
		1.746 (n_E) at 1.0 μm
Titanium oxide (TiO_2)	VIS and IR	2.61 (n_O) at 600 nm
		2.90 (n_E) at 600 nm
Yttrium-aluminum-garnet ($Y_3Al_5O_{12}$)	VIS and IR	1.845 at 500 nm
Yttrium orthovanadate (YVO_4)	VIS and IR	1.993 at 630 nm
Zinc selenide (ZnSe)	VIS and IR	2.489 at 1.0 μm
Zinc sulfide (ZnS)	VIS and IR	2.419 at 500 nm
Germanium (Ge)	IR	4.108 at 2.0 μm
Silicon (Si)	IR	3.432 at 3.0 μm

5.2 UV WINDOWS

Some of the commercially available materials for UV (200–400 nm) windows are UV grade FS, BaF_2, CaF_2, LiF, and MgF_2, which are included in Table 5.2.

The refractive indices of these window materials are obtained from the Sellmeier Equation 5.4.

$$n^2(\lambda) = A + \frac{B_1\lambda^2}{\lambda^2 - C_1^2} + \frac{B_2\lambda^2}{\lambda^2 - C_2^2} + \frac{B_3\lambda^2}{\lambda^2 - C_3^2} \tag{5.4}$$

where λ is the wavelength of light. The constants A, B_1, B_2, B_3, C_1, C_2, and C_3 are listed in Table 5.3 (Refractive Index 2017).

Table 5.4 lists refractive indices of the UV window materials using the Sellmeier Equation 5.4 with the Sellmeier coefficients given in Table 5.3.

TABLE 5.3
Sellmeier Coefficients of the UV Window Materials

Material	A	B_1	B_2	B_3	C_1 (μm)	C_2 (μm)	C_3 (μm)
Fused silica	1.0000	0.696166	0.407943	0.897479	0.068404	0.11624	9.896
BaF_2	1.3397	0.81070	0.19652	4.52469	0.10065	29.87	53.82
CaF_2	1.3397	0.69913	0.11994	4.35181	0.09374	21.18	38.46
LiF	1.0000	0.92549	6.96747	0.00000	0.07376	32.79	0.000
MgF_2	1.2762	0.60967	0.0080	2.14973	0.08636	18.0	25.0

TABLE 5.4
Refractive Indices of UV Window Materials

λ (nm)	FS	BaF_2	CaF_2	LiF	MgF_2
200	1.551	1.557	1.495	1.439	1.423
225	1.524	1.534	1.478	1.427	1.411
250	1.507	1.519	1.467	1.419	1.403
275	1.496	1.509	1.460	1.413	1.397
300	1.488	1.501	1.454	1.409	1.393
320	1.483	1.496	1.451	1.406	1.391
350	1.477	1.491	1.447	1.404	1.388
380	1.473	1.487	1.444	1.400	1.385
400	1.470	1.485	1.442	1.399	1.384

Reflection loss for a plane parallel window is given by

$$R_L = \left[1 + \frac{(1-R)^2 e^{-2\alpha t_w}}{1 - R^2 e^{-2\alpha t_w}} \right] R \tag{5.5}$$

Figure 5.1 shows the reflection loss, assuming α equal to 0, spectra of uncoated FS, BaF$_2$, CaF$_2$, LiF, and MgF$_2$ plane parallel windows for the UV using the refractive index data given in Table 5.4.

There are three grades of synthetic fused silica: optical quality, UV, and IR. The UV grade has high transmittance (85%) starting at 180 nm whereas the optical quality has high transmittance (85%) starting at 260 nm. Figure 5.2 shows the UV external transmittance spectrum of a 10-mm thick uncoated window of UV grade fused silica (Melles Griot 2017).

Absorption loss of the UV grade fused silica is very low in the UV region above 180 nm. AR coatings can reduce the reflection loss of the UV windows to <1.0%. Therefore, UV grade fused silica is one of the best choices for a UV window.

Absorption coefficients of LiF, BaF$_2$, CaF$_2$, and MgF$_2$ are listed in Table 5.5.

Considering the values of the absorption coefficients in Table 5.5, LiF, BaF$_2$, CaF$_2$, and MgF$_2$ are all reasonably good choices for UV windows.

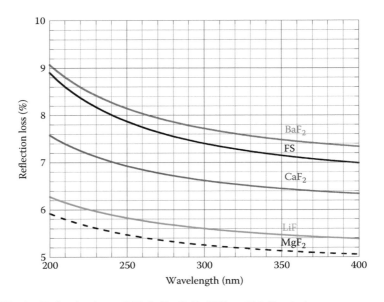

FIGURE 5.1   Reflection loss of FS, BaF$_2$, CaF$_2$, LiF, and MgF$_2$ windows.

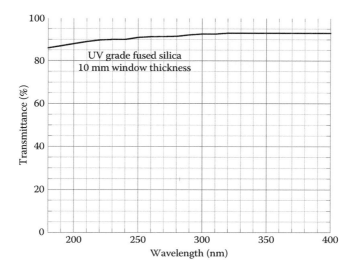

FIGURE 5.2 Transmittance of UV grade fused silica window. (Courtesy of Melles Griot.)

TABLE 5.5
Absorption Coefficients of LiF, BaF$_2$, CaF$_2$, and MgF$_2$

λ (nm)	LiF	BaF$_2$	CaF$_2$	MgF$_2$
200	0.05 cm^{-1}	0.20 cm^{-1}	0.10 cm^{-1}	0.07 cm^{-1}
400	0.02 cm^{-1}	0.08 cm^{-1}	0.01 cm^{-1}	~0.02 cm^{-1}

5.3 VIS AND NIR WINDOWS

Several materials are commercially available for VIS and IR windows. These include BK7, BaF$_2$, CaF$_2$, LiNbO$_3$, MgO, KCl, Al$_2$O$_3$, TiO$_2$, Y$_3$Al$_5$O$_{12}$, YVO$_4$, ZnSe, and ZnS. The refractive indices of these materials are calculated using the Sellmeier Equation 5.4 with the Sellmeier coefficients given in Table 5.6.

TABLE 5.6
Sellmeier Coefficients of VIS and NIR Window Materials

Material	A	B$_1$	B$_2$	B$_3$	C$_1$ (µm)	C$_2$ (µm)	C$_3$ (µm)
BK7	1.0000	1.039612	0.231792	1.010469	0.07746	0.14148	10.176
LiNbO$_3$	1.0000	2.6734	1.2290	12.614	0.13282	0.24319	21.785
Al$_2$O$_3$	1.0000	1.431349	0.650547	5.341402	0.26956	0.34543	4.2460
KBr	1.2649	0.30523	0.41620	0.18870	0.100	0.131	0.162
Y$_3$Al$_5$O$_{12}$	1.0000	2.28200	3.27644	0.0000	0.10886	16.8147	0.0000
ZnSe	1.0000	4.458137	0.467216	2.895663	0.20086	0.39137	47.136

Table 5.7 lists refractive indices of FS, BK7, LiNbO$_3$, Al$_2$O$_3$, KBr, Y$_3$Al$_5$O$_{12}$, and ZnSe VIS and NIR window materials using the Sellmeier Equation 5.6 with the Sellmeier coefficients given in Table 5.6.

Table 5.8 lists the refractive indices of BaF$_2$, CaF$_2$, LiF, and MgF$_2$, NaCl, and KCl window materials.

TABLE 5.7
Refractive Indices of FS, N-BK7, LiNbO$_3$, Al$_2$O$_3$, KBr, Y$_3$Al$_5$O$_{12}$, and ZnSe

λ (nm)	SiO$_2$	N-BK7	LiNbO$_3$	Al$_2$O$_3$	KBr	Y$_3$Al$_5$O$_{12}$	ZnSe
400	1.470	1.531	2.439	2.571	1.511	1.861	4.2317
500	1.462	1.521	2.341	2.062	1.497	1.842	2.743
600	1.458	1.516	2.296	1.919	1.490	1.832	2.614
700	1.455	1.513	2.271	1.846	1.486	1.826	2.557
800	1.453	1.511	2.255	1.797	1.483	1.821	2.524
900	1.452	1.509	2.244	1.758	1.481	1.818	2.503
1000	1.450	1.508	2.236	1.725	1.480	1.816	2.489
1100	1.449	1.506	2.230	1.693	1.479	1.814	2.479
1200	1.448	1.505	2.225	1.660	1.478	1.812	2.471
1300	1.447	1.504	2.220	1.627	1.478	1.811	2.466
1400	1.446	1.503	2.216	1.591	1.477	1.809	2.461
1500	1.445	1.501	2.213	1.551	1.477	1.808	2.457

TABLE 5.8
Refractive Indices of BaF$_2$, CaF$_2$, LiF, and MgF$_2$, NaCl, and KCl

λ (nm)	BaF$_2$	CaF$_2$	CsBr	LiF	MgF$_2$	NaCl	KCl
400	1.485	1.442	1.733	1.399	1.384	1.568	1.511
500	1.478	1.437	1.707	1.394	1.380	1.552	1.497
600	1.474	1.434	1.694	1.392	1.378	1.544	1.490
700	1.472	1.432	1.687	1.390	1.376	1.539	1.486
800	1.471	1.431	1.682	1.389	1.375	1.536	1.483
900	1.469	1.430	1.679	1.388	1.374	1.534	1.481
1000	1.469	1.429	1.676	1.387	1.374	1.532	1.480
1100	1.468	1.428	1.675	1.386	1.373	1.531	1.479
1200	1.468	1.428	1.673	1.386	1.373	1.530	1.478
1300	1.467	1.427	1.673	1.385	1.372	1.530	1.478
1400	1.467	1.427	1.672	1.384	1.372	1.529	1.477
1500	1.466	1.426	1.671	1.383	1.371	1.529	1.477

Figure 5.3 shows the reflection loss spectra of uncoated fused silica, BK7, CaF_2, and KCl windows for VIS and NIR.

Figure 5.4 illustrates the reflection loss spectra of uncoated BaF_2, KBr, LiF, MgF_2, and NaCl windows for VIS and NIR.

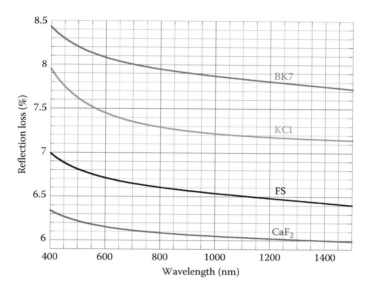

FIGURE 5.3 Reflection loss of uncoated FS, BK7, CaF_2, and KCl windows.

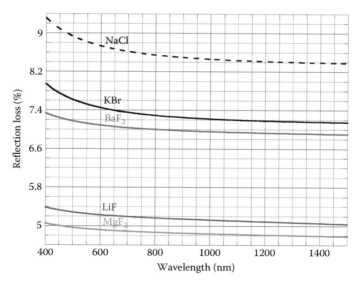

FIGURE 5.4 Reflection loss of BaF_2, KBr, LiF, MgF_2, and NaCl windows.

Figure 5.5 shows VIS and NIR transmittance spectrum of a 10-mm thick uncoated window of BK7, which is one of the common window materials for this spectral range (Schott 2017). It is also an inexpensive window material.

UV grade FS has a strong OH-absorption band in the wavelength range 2600–2800 nm. However, this absorption band is absent in the IR grade fused silica. Figure 5.6 shows the transmittance of a 10-mm thick IR grade FS window (ISP Optics 2017).

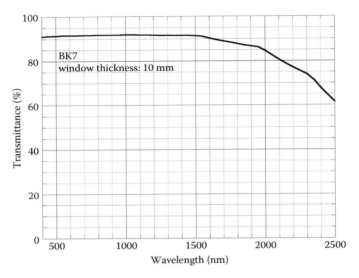

FIGURE 5.5   Transmittance of a 10-mm thick uncoated BK7 window. (Courtesy of Schott.)

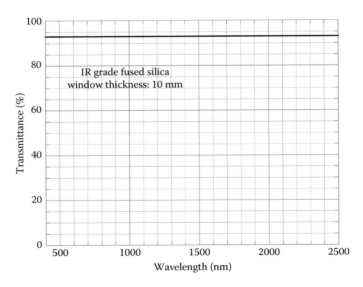

FIGURE 5.6   Transmittance of a 10-mm uncoated IR fused silica window. (Courtesy of ISP Optics.)

5.4   IR WINDOWS

Some of the common materials used for IR windows include BaF_2, CaF_2, CsBr, Ge, KRS-5, NaCl, and ZnSe. Table 5.9 lists IR refractive indices of BaF_2, CaF_2, CsBr, Ge, KRS-5, and ZnSe windows.

Figure 5.7 shows the IR reflection loss of plane parallel windows of BaF_2, CaF_2, CsBr, Ge, KRS-5, and ZnSe.

Table 5.10 lists values of the absorption coefficient α (cm^{-1}) obtained from the data for the internal transmittance of 1-mm thick BaF_2, CaF_2, Ge, and KBr windows (ISP Optics 2017).

External transmittance of uncoated windows of BaF_2, CaF_2, Ge, and KBr was obtained using values of α (cm^{-1}) listed in Table 5.10 and reflection loss for these materials obtained from the refractive index data. Figures 5.8 through 5.11 show the IR transmittance spectrum of 1-mm thick uncoated windows of BaF_2, CaF_2, Ge, and KBr.

TABLE 5.9
IR Refractive Indices of BaF_2, CaF_2, CsBr, Ge, KRS-5, and ZnSe

λ (μm)	BaF_2	CaF_2	CsBr	Ge	KRS-5	ZnSe
2.0	1.465	1.424	1.671		2.395	2.446
2.5	1.463	1.421	1.670	4.065	2.389	2.441
3.0	1.461	1.418	1.670	4.044	2.386	2.438
4.0	1.457	1.410	1.669	4.025	2.382	2.433
5.0	1.451	1.399	1.668	4.016	2.380	2.430
6.0	1.444	1.396	1.667	4.012	2.378	2.426
7.0	1.436	1.369	1.666	4.009	2.376	2.422
8.0	1.426	1.350	1.665	4.007	2.375	2.417
9.0	1.415	1.327	1.664	4.005	2.373	2.412
10.0	1.401	1.300	1.663	4.004	2.371	2.407
11.0	1.387	1.268	1.662	4.004	2.369	2.400
12.0	1.370	1.230	1.660	4.003	2.366	2.393
13.0	1.351		1.659		2.364	2.385
14.0	1.329		1.657		2.361	2.376
15.0	1.305		1.655		2.358	2.367
20.0			1.644		2.341	
25.0			1.629		2.318	
30.0			1.610		2.289	
35.0			1.587		2.253	
40.0			1.559			
45.0			1.524			
50.0			1.483			
55.0			1.432			

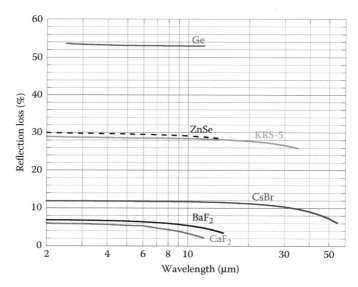

FIGURE 5.7   Reflection loss of BaF$_2$, CaF$_2$, CsBr, Ge, KRS-5, and ZnSe windows.

TABLE 5.10
Absorption Coefficient α (cm⁻¹) for BaF$_2$, CaF$_2$, Ge, and KBr

λ (μm)	BaF$_2$	CaF$_2$	Ge	KBr
1.0	0.30	0.10		0.20
3.0	0.30	0.10	0.30	0.20
5.0	0.30	0.10	0.30	0.20
6.0	0.30	0.20	0.30	0.20
7.0	0.30	0.30	0.30	0.20
8.0	0.30	1.28	0.30	0.20
9.0	0.30	5.28	0.30	0.20
10.0	1.63	16.6	0.41	0.20
12.0	8.68		3.57	0.20
15.0			5.80	0.20
20.0				0.83
30.0				13.5

Source: ISP Optics, www.ispoptics.com, 2017.

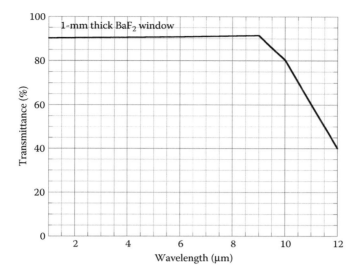

FIGURE 5.8 Transmittance of 1-mm uncoated BaF$_2$ window.

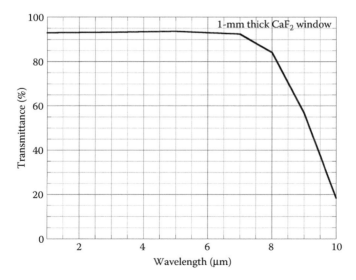

FIGURE 5.9 Transmittance of 1-mm uncoated CaF$_2$ window.

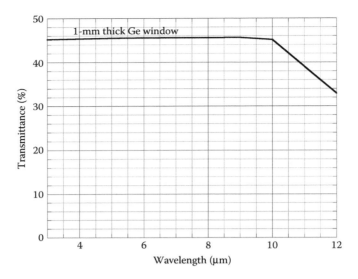

FIGURE 5.10   Transmittance of 1-mm thick uncoated Ge window.

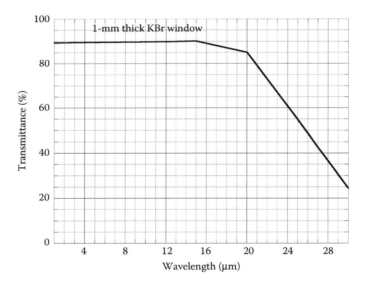

FIGURE 5.11   Transmittance of 1-mm thick uncoated KBr window.

5.5   AR COATING MATERIALS

Refractive indices for materials commonly used for AR coatings are listed in Table 5.11

TABLE 5.11
Refractive Indices of Materials Commonly Used for AR Coatings

Material	n (300 nm)	n (1000 nm)	n (5000 nm)
MgF_2	1.40	1.33	1.294
SiO_2	1.53	1.4598	1.398
Al_2O_3	1.69	1.60	1.578
Ta_2O_5	2.50	2.085	
HfO_2	2.13	1.97	
YF_3	1.59	1.549	1.531
ZnS		2.27	2.245

5.6   HOMEWORK PROBLEMS

5.1 A 50-mm diameter Ge window is subjected to 150 psi pressure differential. Determine its minimum thickness assuming a safety factor of 6.

5.2 A 25-mm diameter CaF_2 window is subjected to 15 psi pressure differential. Determine its minimum thickness assuming a safety factor of 4.

5.3 Calculate the reflection loss of an uncoated Ge window at the CO_2 laser wavelength of 10.6 μm using a value of 4.00 for the refractive index at 10.6 μm.

5.4 Calculate the reflection loss of an uncoated KRS-5 window at the CO_2 laser wavelength of 10.6 μm using a value of 2.37 for the refractive index at 10.6 μm.

5.5 Determine the reflection loss of an uncoated UV grade fused silica window at the He-Ne laser wavelength of 632 nm using a value of 1.457 for the refractive index.

5.6 Calculate the internal and external transmittance of a 2.0-mm thick uncoated Ge window at 12 μm using a value of 3.57 cm^{-1} for the absorption coefficient and 4.003 for the refractive index.

5.7 Determine the internal and external transmittance of a 2.0-mm thick CaF_2 window at 9 μm using a value of 5.28 cm^{-1} for the absorption coefficient and 1.327 for the refractive index.

5.8 Determine the materials that may be used for the 3–12 μm AR coatings of Ge windows. Assume the refractive index of Ge to be 4.00.

5.9 Determine the materials that may be used for the 532 nm AR coating of a CaF_2 window. Assume the refractive index of CaF_2 to be 1.435.

5.10 Determine the materials that may be used for the 532 nm AR coating of an Al_2O_3 window. Assume the refractive index of Al_2O_3 to be 1.772.

5.11 Deduce an expression for the parallel displacement of a beam of light as it goes through a window, in terms of the angle of incidence, refractive index, and thickness of the window.

6 Optical Filters

6.1 INTRODUCTION

An optical filter is required for defining the wavelength range of the light incident upon an object, which includes the entrance slit of an optical system. There are several types of optical filters: (1) colored glass filters, (2) dielectric filters, (3) neutral density filters, and (4) Raman filters (used in Raman spectroscopy).

6.2 COLORED GLASS FILTERS

Colored glass filters are manufactured using different types of Schott color glass. Bandpass and longpass colored glass filters for UV, VIS, and IR wavelengths are commercially available from several vendors including Edmund Optics, Newport, Schott, and Thorlabs. A bandpass filter has a transmission band surrounded by two blocking bands. A longpass filter is characterized by its cut-on wavelength. Table 6.1 has descriptions of different types of Schott color glasses (Schott 2017).

Figure 6.1 shows the external transmittance spectrum of a 1.0-mm thick color glass bandpass filter BG60, which has two bandpasses with peak transmittance at 500 and 2500 nm, respectively. The external transmittance is equal to the internal times (1-reflection loss).

Figure 6.2 shows the external transmittance spectrum of a 1.0-mm thick color glass longpass filter GG395 with cut-on wavelength of 395 nm. The cut-on wavelength of a longpass filter denotes the wavelength at which the transmittance increases to 50%. Similarly, the cut-off wavelength of a shortpass filter denotes the wavelength at which the transmittance decreases to 50%.

Some of the commercially available Schott longpass and bandpass filters are listed in Tables 6.2 and 6.3, respectively.

TABLE 6.1
Schott Color Glasses

Type	Description
UG	UV transmitting black and blue glasses
BG	Blue, blue-green, and multi-band glasses
VG	Green glasses
GG	IR transmitting virtually colorless to yellow glasses
OG	IR transmitting orange glasses
RG	IR transmitting red and black glasses
NG	Neutral density glasses with uniform attenuation in the visible range
N-WG	Colorless glasses with different cut offs in the UV, transmitting in the VIS and IR
KG	Virtually colorless glasses with high transmission in the VIS and absorption in the IR (heat protection filters)

Source: Schott, www.us.schott.com, 2017.

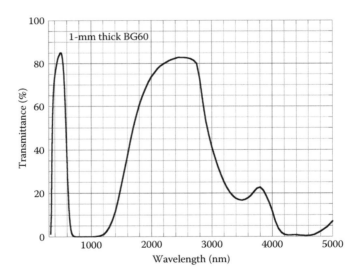

FIGURE 6.1   Transmittance of a 1.0-mm thick BG60 filter. (Courtesy of Schott.)

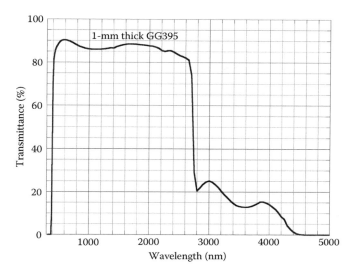

FIGURE 6.2 Transmittance of a 1.0-mm thick color glass longpass filter GG395. (Courtesy of Schott.)

TABLE 6.2
Commercially Available Schott
Longpass Filters

Filter	Cut-On λ (nm)
N-WG280	280 ± 6
N-WG295	295 ± 6
N-WG305	305 ± 6
N-WG320	320 ± 6
GG395	395 ± 6
GG400	400 ± 6
GG420	420 ± 6
GG435	435 ± 6
GG455	455 ± 6
GG475	475 ± 6
GG495	495 ± 6
OG515	515 ± 6
OG530	530 ± 6
OG550	550 ± 6
OG570	570 ± 6
	(Continued)

TABLE 6.2 (*Continued*)
**Commercially Available Schott
Longpass Filters**

Filter	Cut-On λ (nm)
OG590	590 ± 6
RG610	610 ± 6
RG630	630 ± 6
RG645	645 ± 6
RG665	665 ± 6
RG695	695 ± 6
RG715	715 ± 6
RG780	780 ± 6
RG830	830 ± 6
RG850	850 ± 6
RG900	900 ± 6
RG1000	1000 ± 6

Source: Schott, www.us.schott.com, 2017.

TABLE 6.3
**Commercially Available Bandpass
Filters**

Filter	λ (nm)
UG5	240–395
UG11	275–375
KG2	304–785
BG3	315–445
KG3	315–710
KG1	315–750
UG1	325–385
KG5	330–665
BG40	335–610
BG39	360–580
BG18	412–569
BG7	435–500
VG9	485–565
BG36	900–1375

Source: Schott, www.us.schott.com, 2017.

6.3 DIELECTRIC FILTERS

Dielectric filters are fabricated by the deposition of dielectric films on substrates, which are transparent to the wavelength region of interest. Bandpass, longpass, and shortpass dielectric filters are commercially available from several vendors including Edmund Optics (Edmund Optics 2017), Newport (Newport 2017), and Thorlabs (Thorlabs 2017).

Figure 6.3 shows the transmittance spectrum of a dielectric bandpass filter with center wavelength (CWL) of 500 ± 2 nm and full-width at half-maximum (FWHM) of 10 ± 2 nm (Thorlabs 2017).

Table 6.4 lists some of the commercially available UV/VIS dielectric bandpass filters.

Figure 6.4 shows the transmittance spectrum of a 532 ± 0.2 nm dielectric laser line filter.

Table 6.5 lists some of the commercially available UV/VIS dielectric laser line filters.

Figures 6.5 through 6.7 each show the transmittance spectra of 780 ± 2, 1064±2, and 1650 ± 2.4 nm NIR dielectric bandpass filters, respectively.

Table 6.6 has some of the commercially available NIR dielectric bandpass filters in the wavelength range 700–1500 (Thorlabs 2017).

Figure 6.8 shows the transmittance spectrum of a 780 ± 2 nm NIR diode laser line dielectric filter (Thorlabs 2017).

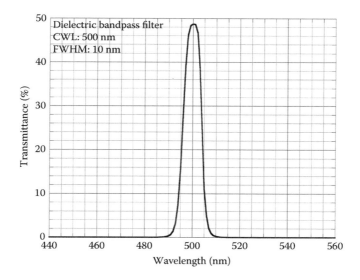

FIGURE 6.3   Transmittance of a 500-nm dielectric bandpass filter. (Courtesy of Thorlabs.)

TABLE 6.4
Commercially Available UV/VIS Dielectric Bandpass Filters

CWL (nm)	FWHM (nm)	Minimum Peak Transmittance (%)
340 ± 2, 350 ± 2, 360 ± 2, 370 ± 2, 380 ± 2, 390 ± 2, 400 ± 2	10 ± 2	25
400 ± 8	40 ± 8	45
405 ± 2, 410 ± 2, 420 ± 2, 430 ± 2, 440 ± 2, 450 ± 2	10 ± 2	37
450 ± 8	40 ± 8	45
460 ± 2, 470 ± 2, 480 ± 2, 490 ± 2, 500 ± 2	10 ± 2	45
500 ± 8	40 ± 8	70
510 ± 2, 520 ± 2, 530 ± 2, 540 ± 2, 550 ± 2, 560 ± 2, 570 ± 2, 580 ± 2, 590 ± 2	10 ± 2	50
550 ± 8	40 ± 8	70
610 ± 2, 610 ± 2, 620 ± 2, 630 ± 2, 640 ± 2, 650 ± 2, 660 ± 2, 670 ± 2, 680 ± 2, 690 ± 2	10 ± 2	50
650 ± 8	40 ± 8	70

Source: Thorlabs, www.thorlabs.com, 2017.

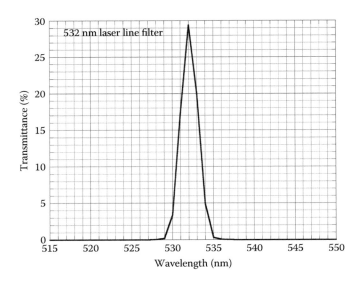

FIGURE 6.4   Transmittance of a 532-nm dielectric laser line filter. (Courtesy of Thorlabs.)

TABLE 6.5
Commercially Available UV/VIS Dielectric Laser Line Filters

CWL (nm)	FWHM (nm)	Minimum Peak Transmittance (%)
355 ± 2, 441.6 ± 2, 457.9 ± 2	10 ± 2	25
480 ± 0.2	1 ± 0.2	40
488 ± 0.6	3 ± 0.6	45
488 ± 2	10 ± 2	65
514.5 ± 0.2	1 ± 0.2	40
514.5 ± 0.6	3 ± 0.6	55
514 ± 2	10 ± 2	65
532 ± 0.2	1 ± 0.2	40
532 ± 0.6	3 ± 0.6	60
532 ± 2	10 ± 2	70
543.5 ± 2	10 ± 2	70
632.8 ± 0.2	1 ± 0.2	50
632.8 ± 0.6	3 ± 0.6	65
632.8 ± 2	10 ± 2	70
647.1 ± 2	10 ± 2	70
694.3 ± 2	10 ± 2	70

Source: Thorlabs, www.thorlabs.com, 2017.

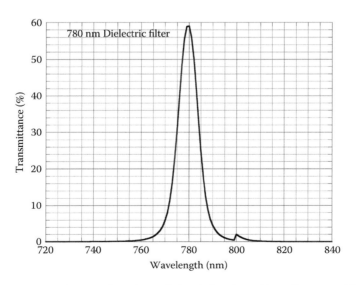

FIGURE 6.5   Transmittance of a 780-nm dielectric bandpass filter. (Courtesy of Thorlabs.)

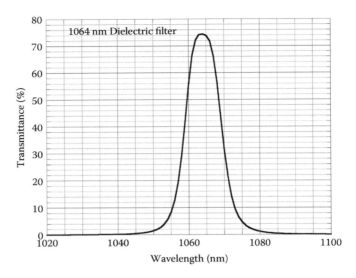

FIGURE 6.6   Transmittance of a 1064-nm dielectric bandpass filter. (Courtesy of Thorlabs.)

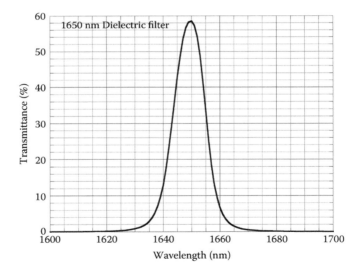

FIGURE 6.7   Transmittance of a 1650-nm dielectric bandpass filter. (Courtesy of Thorlabs.)

TABLE 6.6
Commercially Available NIR Dielectric Bandpass Filters

CWL (nm)	FWHM (nm)	Minimum Peak Transmittance) (%)
700 ± 2, 710 ± 2, 720 ± 2, 730 ± 2, 740 ± 2, 750 ± 2	10 ± 2	50
760 ± 2, 770 ± 2, 780 ± 2, 790 ± 2, 800 ± 2, 810 ± 2	10 ± 2	50
820 ± 2, 830 ± 2, 840 ± 2, 850 ± 2, 860 ± 2, 870 ± 2	10 ± 2	50
880 ± 2, 890 ± 2, 900 ± 2, 910 ± 2, 920 ± 2, 930 ± 2	10 ± 2	50
940 ± 2, 950 ± 2, 960 ± 2, 970 ± 2, 980 ± 2, 990 ± 2	10 ± 2	50
1000 ± 2, 1050 ± 2	10 ± 2	45
1100 ± 2, 1150 ± 2, 1200 ± 2, 1250 ± 2	10 ± 2	40
1300 ± 2.4, 1350 ± 2.4, 1400 ± 2.4, 1450 ± 2.4	12.4 ± 2.4	35

Source: Thorlabs, www.thorlabs.com, 2017.

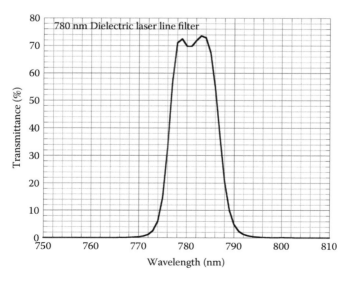

FIGURE 6.8 Transmittance of a 780 ± 2 nm dielectric laser line filter. (Courtesy of Thorlabs.)

Table 6.7 lists some of the commercially available NIR laser line dielectric filters (Thorlabs 2017).

Figure 6.9 shows the transmittance spectrum of an IR dielectric bandpass filter (Table 6.8) with 6000 nm CWL and 500 ± 100 nm FWHM (Thorlabs 2017).

A dielectric notch filter is used to block the laser light. In Raman spectroscopy, it is important to block the pump laser light. Figure 6.10 shows the transmittance spectrum of a 594-nm dielectric notch filter (Thorlabs 2017).

TABLE 6.7

Commercially Available NIR Laser Line Dielectric Filters

CWL (nm)	FWHM (nm)	Minimum Peak Transmittance (%)
730 ± 2, 780 ± 2	10 ± 2	70
830 ± 2, 850 ± 2, 880 ± 2	10 ± 2	70
905 ± 2	10 ± 2	70
1064 ± 0.6	3 ± 0.6	55
1064 ± 2	10 ± 2	70
1152 ± 2	10 ± 2	45
1550 ± 2.4	12 ± 2.4	50

Source: Thorlabs, www.thorlabs.com, 2017.

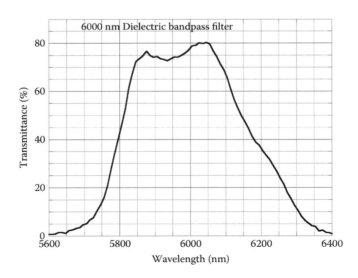

FIGURE 6.9 Transmittance of a 6000-nm dielectric bandpass filter.

TABLE 6.8

Commercially Available IR Dielectric Bandpass Filters

CWL (nm)	FWHM (nm)	Minimum Peak Transmittance (%)
1750	500 ± 100	70
2000, 2250, 2500, 2750	500 ± 100	70
3000, 3250, 3500, 3750	500 ± 100	70
4000, 4250, 4500, 4750	500 ± 100	70
5000, 5250, 5500, 5750	500 ± 100	70
6000	500 ± 100	70

Source: Thorlabs, www.thorlabs.com, 2017.

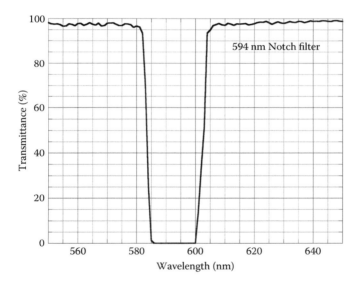

FIGURE 6.10 Transmittance of a 594-nm dielectric notch filter. (Courtesy of Thorlabs.)

Table 6.9 lists several commercially available dielectric notch filters.

Dichroic longpass and shortpass filters are designed for a 45° angle of incidence. The reflected light is rejected. Tables 6.10 and 6.11 list some of the commercially available dichroic longpass and shortpass filters, respectively (Edmund Optics 2017).

TABLE 6.9
Commercially Available Dielectric Notch Filters

CWL (nm)	FWHM (nm)	Passband ($T_{ave} > 90\%$) (nm)
405 ± 2	13	360–390; 420–570
488 ± 2	15	400–471; 504–650
514 ± 2	17	400–496; 532–690
533 ± 2	17	400–517; 548–710
561 ± 2	18	425–542; 580–740
594 ± 2	23	440–572; 616–810
633 ± 2	25	475–613; 653–900
658 ± 2	26	500–634; 682–900
785 ± 2	33	590–760; 810–1040
808 ± 2	34	610–778; 838–1060
980 ± 2	41	700–948; 1012–1300
1064 ± 2	44	800–1031; 1097–1400

Source: Thorlabs, www.thorlabs.com, 2017.

TABLE 6.10
Commercially Available Dichroic Longpass Filters

Cut-On λ (nm)	Transmitted Band (nm)	Reflected Band (nm)
400	420–1600	350–375
450	470–1600	350–430
500	520–1600	350–480
550	575–1600	415–515
600	625–1600	460–570
650	675–1600	495–610
700	725–1600	535–660
750	780–1600	565–715
800	830–1600	600–760
850	880–1600	635–805
900	935–1600	675–855
950	985–1600	715–900
1000	1035–1600	750–950
1050	1085–1600	790–1000
1100	1135–1600	825–1045
1150	1190–1600	875–1095
1200	1240–1600	920–1160

Source: Edmund Optics, www.edmundoptics.com/optics, 2017.

TABLE 6.11
Commercially Available Dichroic Shortpass Filters

Cut-On λ (nm)	Transmitted Band (nm)	Reflected Band (nm)
400	325–385	420–485
450	325–430	470–545
500	325–480	520–610
550	400–530	575–725
600	400–580	625–795
650	400–630	675–850
700	400–680	725–900
750	400–725	800–990
800	400–775	850–1050
850	400–820	910–1110
900	465–865	960–1170
950	495–912	997–1235
1000	520–960	1050–1300
1050	546–1008	1102–1365
1100	572–1056	1155–1430
1150	598–1104	1207–1495
1200	624–1152	1260–1560

Source: Edmund Optics, www.edmundoptics.com/optics, 2017.

6.4 NEUTRAL DENSITY FILTERS

Transmittance of a neutral density (ND) filter should remain constant over a given wavelength range. Transmittance of an ND filter is given by

$$T = 10^{-OD} \tag{6.1}$$

where OD is the optical density of the ND filter. ND filters are either absorptive or reflective. Absorptive ND filters are fabricated from Schott glass and result from OD ranging from 0.1 to 8.0. Figure 6.11 shows the transmittance spectrum of an absorptive 3.0 ND filter for the wavelength range 450–650 nm (Thorlabs 2017). Transmittance of the 3.0 ND filter should be 0.10%, which is close to the values of T that vary between 0.06 and 0.12 in the spectrum of Figure 6.11.

Figure 6.12 shows the transmittance spectrum of a 3.0 ND reflective ND filter for the wavelength range 450–650 nm (Thorlabs 2017). The measured values of T in the spectrum of Figure 6.12 vary between 0.22% at 450 nm and 0.19% at 650 nm compared with the expected value of 0.10.

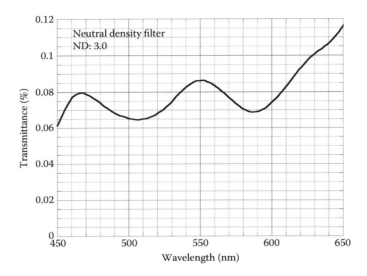

FIGURE 6.11 Transmittance of an absorptive 3.0 ND filter. (Courtesy of Thorlabs.)

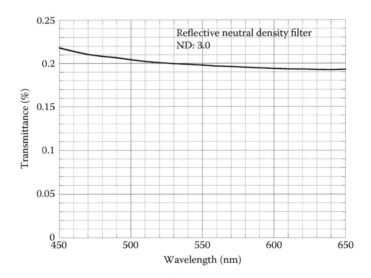

FIGURE 6.12 Transmittance of a reflective 3.0 ND filter. (Courtesy of Thorlabs 2016.)

6.5 RAMAN FILTERS

Raman filters are used to suppress the pump laser line in Raman spectroscopy. Raman filters are either longpass or shortpass filters with sharp turn-on or turn-off. Semrock is the world's leading supplier of high-quality Raman filters. Figure 6.13 shows the transmittance of a 532 nm RazorEdge longpass Raman filter (Semrock 2017).

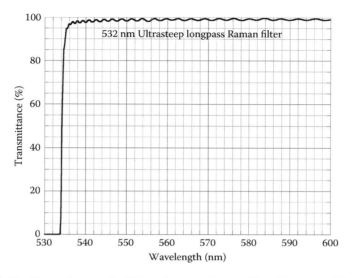

FIGURE 6.13   Transmittance of a 532-nm longpass Raman filter. (Courtesy of Semrock.)

The transmittance increases from $1.0 \times 10^{-4}\%$ at 532 nm laser wavelength to 70.3% at 534.4 nm. This corresponds to a transition width of <84 cm^{-1}. There is no observable temperature shift of the cut-on wavelength in the operating temperature range -47–$71°C$ (Semrock 2017).

Figure 6.14 shows the transmittance spectrum of a RazorEdge shortpass Raman filter (Semrock 2017). The transmittance increases from $9.0 \times 10^{-7}\%$ at 532 nm

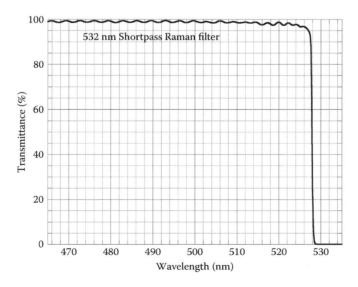

FIGURE 6.14   Transmittance of a 532-nm shortpass Raman filter. (Courtesy of Semrock.)

laser wavelength to 63.3% at 527.8 nm. This corresponds to a transition width of $<150 \text{ cm}^{-1}$. Again, there is no observable temperature shift in the cut-off wavelength in the operating range -47–$71°C$ (Semrock 2017).

6.6 HOMEWORK PROBLEMS

6.1 Calculate and plot (1) external transmittance and (2) optical density (OD) of a 1-mm thick BG60 Schott glass filter as a function of the wavelength using the following internal transmittance data. Assume the refractive index to be 1.5.

λ (nm)	350	400	450	500	550	600
T_i	0.352	0.809	0.896	0.932	0.821	0.390
λ (nm)	650	700	750	800	850	900
T_i	6.4×10^{-2}	4.3×10^{-3}	2.8×10^{-4}	4.6×10^{-5}	2.5×10^{-5}	3.6×10^{-5}

6.2 Calculate OD of a 0.1-mm thick BG60 Scott glass filter at 600 nm.

6.3 Calculate and plot the external transmittance of a 3.0-mm thick Schott OG530 longpass filter using the following internal transmittance data. The refractive index of OG530 is 1.51.

λ (nm)	500	510	520	530	540	550
T_i	$<1 \times 10^{-5}$	5.2×10^{-3}	0.136	0.514	0.802	0.919
λ (nm)	560	570	580	590	600	610
T_i	0.961	0.976	0.982	0.985	0.986	0.986

6.4 Calculate and plot the external transmittance of a 1.0-mm thick Schott UG1 bandpass filter using the following internal transmittance data. The refractive index of UG1 is 1.57

λ (nm)	290	300	310	320	330	340	350
T_i	3.7×10^{-2}	0.155	0.335	0.519	0.659	0.751	0.807
λ (nm)	360	370	380	390	400	410	420
T_i	0.833	0.812	0.707	0.438	0.138	1.7×10^{-2}	8.6×10^{-4}

6.5 Calculate the OD of a Thorlabs dielectric bandpass filter with center wavelength (CWL) of 500 ± 2 nm and full-width at half-maximum (FWHM) of 10 ± 2 nm using the following data for external transmittance.

λ (nm)	490	491	492	493	494	495	496
T_E	0.003	0.007	0.014	0.032	0.075	0.158	0.277
λ (nm)	497	498	499	500	501	502	503
T_E	0.392	0.461	0.484	0.487	0.486	0.466	0.386
λ (nm)	504	505	506	507	508	509	510
T_E	0.251	0.112	0.051	0.021	0.009	0.005	0.002

6.6 Show that a proper combination of longpass and shortpass filters can lead to a bandpass filter.

6.7 Calculate and plot OD versus wavelength of a Thorlabs 594 ± 2 nm notch filter using the data given below:

λ (nm)	581	582	583	584	585	586	587
T_E	0.96237	0.93288	0.69963	0.25154	0.013	0.00001	0.00001
λ (nm)	588	589	590	591	592	593	594
T_E	0.00001	0.00001	0.00001	0.00001	0.00001	0.00001	0.00001
λ (nm)	595	596	597	598	599	600	601
T_E	0.00001	0.00001	0.00001	0.00001	0.00001	0.00011	0.13816
λ (nm)	602	603	604	605	606	607	608
T_E	0.33374	0.51804	0.93445	0.94894	0.96873	0.97385	0.97866

6.8 Calculate and plot OD versus wavelength of Semrock 532 nm longpass Raman filter using the data given below:

λ (nm)	532.0	532.2	532.4	532.6	532.8	533.0	533.2
T_E	9.9×10^{-7}	5.5×10^{-6}	4.9×10^{-5}	5.8×10^{-5}	6.8×10^{-5}	9.6×10^{-5}	1.9×10^{-4}
λ (nm)	533.4	533.6	533.8	534.0	534.2	534.4	534.6
T_E	3.3×10^{-4}	6.2×10^{-4}	4.0×10^{-3}	8.5×10^{-2}	0.538	0.703	0.847
λ (nm)	534.8	535.0	535.2	535.4	535.6	535.8	536.0
T_E	0.868	0.912	0.947	0.948	0.954	0.968	0.969
λ (nm)	536.2	536.4	536.6	536.8	537.0	537.2	537.4
T_E	0.966	0.970	0.977	0.980	0.976	0.972	0.972

6.9 Calculate and plot OD versus wavelength of Semrock 532 nm shortpass Raman filter using the data given below:

λ (nm)	525.0	525.2	525.4	525.6	525.8	526.0	526.2
T_E	0.968	0.967	0.968	0.969	0.969	0.968	0.966
λ (nm)	526.4	526.6	526.8	527.0	527.2	527.4	527.6
T_E	0.963	0.960	0.956	0.951	0.943	0.931	0.879
λ (nm)	527.8	528.0	528.2	528.4	528.6	528.8	529.0
T_E	0.633	0.259	0.062	0.014	0.004	1.6×10^{-3}	8.4×10^{-4}
λ (nm)	529.2	529.4	529.6	529.8	530.0	530.2	530.4
T_E	5.9×10^{-4}	4.1×10^{-4}	6.8×10^{-5}	9.9×10^{-6}	3.6×10^{-6}	1.5×10^{-6}	8.5×10^{-7}
λ (nm)	530.6	530.8	531.0	531.2	531.4	531.6	531.8
T_E	4.8×10^{-7}	3.7×10^{-7}	3.0×10^{-7}	2.3×10^{-7}	1.6×10^{-7}	9.5×10^{-8}	2.9×10^{-8}

7 Beamsplitters

7.1 INTRODUCTION

Beamsplitters split incident light into two separate beams. Alternatively, beamsplitters can be used in reverse to combine two different beams into a single beam. There are four configurations of the beamsplitters:

1. Standard beamsplitters, which are designed to split incident light at a specific reflection/transmission (R/T) ratio that is independent of the polarization state.
2. Polarizing beamsplitters, which are designed to split unpolarized light into reflected S-polarized light and transmitted P-polarized light.
3. Non-polarizing beamsplitters split incident light into a specific R/T ratio while maintaining the incident light's original polarization state.
4. Dichroic beamsplitters split incident light as a function of wavelength.

There are three types of beamsplitters: (1) plate, (2) cube, and (3) pellicle. A plate beamsplitter consists of a thin plate, which is coated on the front surface. Plate beamsplitters include dichroic beamsplitters and Polka dot beamsplitters, which are fabricated by depositing squares of aluminum coating on a substrate. Polka-dot beamsplitters have a constant R/T ratio over a large spectral range. Cube beamsplitters consist of two right angle prisms that are cemented together; the hypotenuse of one prism is coated. Pellicle beamsplitters are very thin (~2 μm) nitrocellulose membranes, which are bonded to lapped aluminum frames.

7.2 PLATE BEAMSPLITTERS

Most often plate beamsplitters are designed for a 45° angle of incidence (AOI), as shown in Figure 7.1. The second surface of the plate is AR coated to avoid ghosting due to reflection from the second surface. Sometimes the second surface of the plate is wedged to prevent ghosting.

The displacement δ of the transmitted beam from the incident beam is given by

$$\delta = \frac{t}{\sqrt{2}}\left(1 - \tan\theta_t\right) \tag{7.1}$$

where:
    t is thickness of the plate
    θ_t is the angle of transmission that is given by

$$\sin\theta_t = \frac{1}{n\sqrt{2}} \tag{7.2}$$

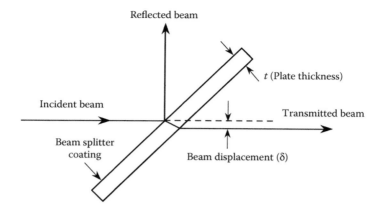

FIGURE 7.1   Plate beamsplitter for a 45° angle of incidence.

where n is the refractive index of the beamsplitter plate. For example, δ is equal to $0.347t$ for n equal to 1.5.

7.2.1   NON-POLARIZING BEAMSPLITTERS

Non-polarizing beamsplitters are available from several vendors including Edmund Optics, Newport, and Thorlabs for the UV, VIS, NIR, and IR wavelength regions with reflection/transmission ratios of 25/75, 30/70, 40/60, 50/50, 70/30, and 75/25. Figures 7.2 and 7.3 show the transmittance and reflectance spectra of a commercially

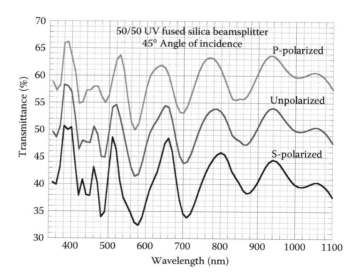

FIGURE 7.2   Transmittance of a UV fused silica beamsplitter. (Courtesy of Thorlabs.)

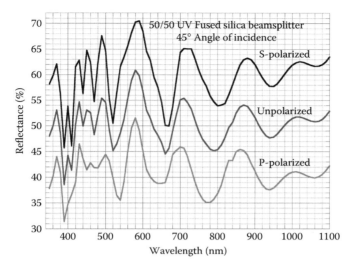

FIGURE 7.3 Reflectance of a UV fused silica beamsplitter. (Courtesy of Thorlabs.)

available 45° AOI, 50/50 non-polarizing beamsplitter of UV-grade fused silica (UVFS) coated for the wavelength region 350–1100 nm (Thorlabs 2017).

Figures 7.4 and 7.5 show the transmittance and reflectance spectra of a 45° AOI IR fused silica broadband beamsplitter coated for the wavelength region 900–2600 nm (Thorlabs 2017).

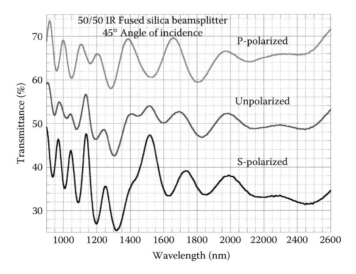

FIGURE 7.4 Transmittance of an IR fused silica non-polarizing beamsplitter. (Courtesy of Thorlabs.)

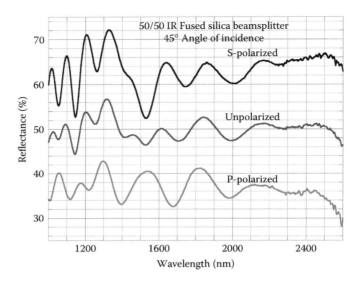

FIGURE 7.5   Reflectance of an IR fused silica non-polarizing beamsplitter. (Courtesy of Thorlabs.)

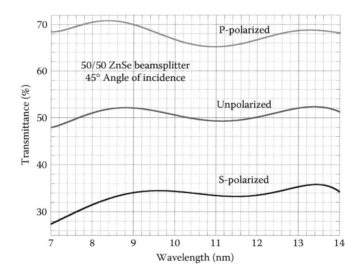

FIGURE 7.6   Transmittance of a ZnSe beamsplitter. (Courtesy of Thorlabs.)

Figure 7.6 shows the transmittance spectrum of a 45° AOI ZnSe broadband non-polarizing beamsplitter coated for the wavelength region 7–14 μm (Thorlabs 2017).

Table 7.1 lists some of the commercially available 45° AOI non-polarizing plate beamsplitters (Thorlabs 2017).

TABLE 7.1
Commercially Available 45° AOI
Non-Polarizing Plate Beamsplitters

Material	λ (nm)
UV FS	250–450
	350–1,100
	400–700
	600–1,700
	700–1,100
	1,200–1,600
IR FS	900–2,600
CaF$_2$	1,000–8,000
ZnSe	7,000–14,000

Source: Thorlabs, www.thorlabs.com, 2017.

7.2.2 Dichroic Beamsplitters

Figure 7.7 shows the transmittance/reflectance spectra of a longpass dichroic beamsplitter with a cut-off wavelength of 550 nm (Thorlabs 2017).

Table 7.2 lists some of the commercially available longpass dichroic beam splitters (Thorlabs 2017).

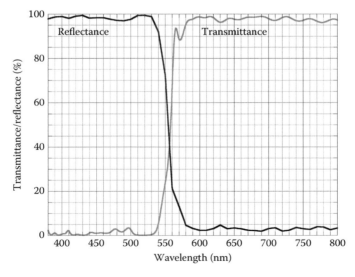

FIGURE 7.7 Transmittance/reflectance of 550-nm longpass dichroic beamsplitter. (Courtesy of Thorlabs.)

TABLE 7.2
Commercially Available Dichroic Beamsplitters

Cut-Off λ (nm)	Transmission Band (nm)	Reflection Band (nm)
425	440–700	380–410
490	510–800	380–475
505	520–700	380–490
550	565–800	380–533
567	584–700	380–550
605	620–700	470–590
638	655–700	580–621
650	685–1600	400–633
900	932–1300	400–862
950	990–1600	420–900
1180	1260–1700	750–1100
1800	1850–2100	1500–1750

Source: Thorlabs, www.thorlabs.com, 2017.

Figure 7.8 shows the transmittance/reflectance spectra of a shortpass dichroic beamsplitter with a cutoff wavelength of 805 nm (Thorlabs 2017).

Table 7.3 lists some of the commercially available shortpass dichroic beam splitters (Thorlabs 2017).

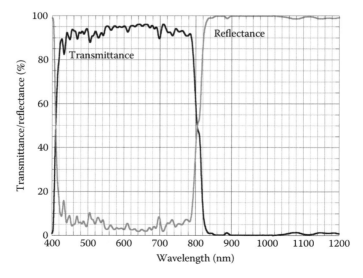

FIGURE 7.8   Transmittance/reflectance of an 805-nm shortpass dichroic beamsplitter. (Courtesy of Thorlabs.)

TABLE 7.3
Commercially Available Shortpass Dichroic Beamsplitters

Cut-Off λ (nm)	Transmission Band (nm)	Reflection Band (nm)
650	400–633	685–1600
805	400–788	823–1300
950	420–900	990–1600
1000	520–985	1020–1550
1180	750–1100	1260–1700
1500	1000–1450	1550–2000

Source: Thorlabs, www.thorlabs.com, 2017.

7.2.3 POLKA DOT BEAMSPLITTERS

A polka dot beamsplitter is fabricated by depositing square dots $(100 \times 100\ \mu m^2)$ of aluminum coating on a substrate. These beamsplitters have a fairly constant R/T ratio over a large spectral range with R/T ratios of 30/70, 50/50, and 70/30. Figure 7.9 shows the reflectance and transmittance spectra of a 45° AOI, 50/50 CaF_2 polka dot beamsplitter for the wavelength range 1000–8000 nm (Thorlabs 2017).

Table 7.4 lists some of the commercially available 45° AOI polka dot beamsplitters (Thorlabs 2017).

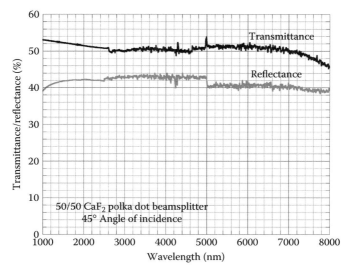

FIGURE 7.9 Transmittance and reflectance of a CaF_2 polka dot beamsplitter. (Courtesy of Thorlabs.)

TABLE 7.4
Commercially Available 45° AOI
Polka Dot Beamsplitters

Substrate Material	λ (nm)
UV FS	250–2,000
B270 Glass	350–2,000
CaF$_2$	180–8,000
ZnSe	2,000–11,000

Source: Thorlabs, www.thorlabs.com, 2017.

7.3 CUBE BEAMSPLITTERS

Cube beamsplitters are fabricated using two right angle prisms. The hypotenuse of one prism is coated and then the two prisms are cemented together to form a cube. Standard cube beamsplitters with R/T ratios of 30/70, 50/50, and 70/30 are commercially available from vendors including Edmund Optics, Newport, and Thorlabs. Non-polarizing and polarizing cube beamsplitters with R/T ratio of 50/50 are also commercially available. For laser applications, the damage threshold of the cement is about 150 W/cm for the Thorlabs beam splitters (Thorlabs 2017).

7.3.1 Non-Polarizing Cube Beamsplitters

Non-polarizing cube beamsplitters are available with broadband AR coatings and R/T ratios of 10/90, 30/70, 50/50, 70/30, and 90/10. Figure 7.10 shows the transmittance

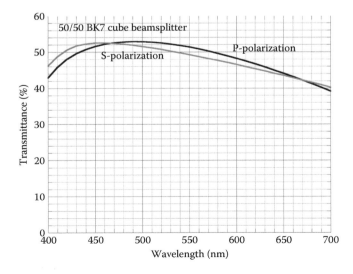

FIGURE 7.10 Transmittance of a BK7 cube beamsplitter. (Courtesy of Thorlabs.)

TABLE 7.5

Commercially Available Non-Polarizing, BK7 Cube Beamsplitters

λ (nm)	R/T
400–700	10/90, 30/70, 50/50, 70/30, 90/10
700–1100	30/70, 50/50, 70/30, 90/10
1100–1600	50/50, 70/30, 90/10

Source: Thorlabs, www.thorlabs.com, 2017.

spectra for P and S polarizations of a 50/50 cube beamsplitter of BK7 for the wavelength range 400–700 nm (Thorlabs 2017).

Table 7.5 lists some of the commercially available non-polarizing, BK7 cube beamsplitters with *R/T* ratios of 10/90, 30/70, 50/50, 70/30, and 90/10 (Thorlabs 2017).

7.3.2 POLARIZING CUBE BEAMSPLITTERS

Polarizing cube beamsplitters split randomly polarized light into two orthogonal, linearly polarized components. The S-polarized component is reflected, while the P-polarized component is transmitted. These beamsplitters are designed to make the most use of the transmitted beam, which has an extinction ratio of > 1000:1. The reflected beam has an extinction ratio of only about 20–100. Broadband polarizing cube beamsplitters of N-SF1 are available for the wavelength ranges 420–680, 620–1000, 900–1300, and 1200–1600 nm with transmittance $T_P > 90\%$ and reflectance $R_S > 99\%$. Laser line cube beamsplitters are available for laser wavelengths of 355, 488, 532, 632, 780, 850, 980, and 1064 nm with $T_P > 95\%$ at the design wavelength.

7.4 PELLICLE BEAMSPLITTERS

Pellicle beamsplitters are made of very thin (2 μm) nitrocellulose film bonded to lapped aluminum frames. Uncoated pellicle beamsplitters have an *R/T* ratio of 8/92. Dielectric-coated pellicle beamsplitters are available with *R/T* ratios of 33/67, 45/55, and 50/50. Figure 7.11 shows the transmittance and reflectance spectra for unpolarized light of a 45° AOI, coated for the wavelength range 400–700 nm with *R/T* of 45/55 (Thorlabs 2017).

The transmittance and reflectance spectra in Figure 7.11 show interference fringes because of the very small (2-μm) thickness of the beamsplitter. The fringe separation is given by

$$\Delta v \, (\text{cm}^{-1}) = \frac{\cos\theta_t}{2nt(\text{cm})} \tag{7.3}$$

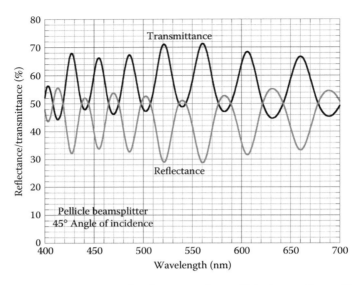

FIGURE 7.11 Transmittance and reflectance of a pellicle beamsplitter. (Courtesy of Thorlabs.)

TABLE 7.6
Commercially Available Pellicle Beamsplitters

R/T	λ (nm)
8/92	400−2400
45/55	300−400
45/55	400−700
45/55	700−900
45/55	1000−2000
45/55	3000−5000

Source: Thorlabs, www.thorlabs.com, 2017.

The fringe spacing in the spectra of Figure 7.11 is consistent with Equation 7.3 using values of 1.5 for n, 2×10^{-4} cm for t, and 27° for θ_r. Table 7.6 lists some of the commercially available pellicle beamsplitters (Thorlabs 2017).

7.5   HOMEWORK PROBLEMS

7.1 A 45° AOI 1-mm thick beamsplitter is made of germanium, which has a refractive index of 4.0. Determine the displacement δ of the transmitted beam.

7.2 Calculate the displacement δ of the transmitted beam as a function of the angle of incidence θ_i of 30°, 40°, 50°, and 60°. The refractive index n of the beamsplitter is equal to 1.5.

7.3 Plot the transmittance spectrum of Figure 7.11 as a function of ν (cm⁻¹) using this chapter's tables with Thorlabs data (Thorlabs 2017). Determine the fringe spacing in the plotted spectrum.

8 Light Sources

8.1 INTRODUCTION

Light sources are used for lighting, spectroscopy, and other technical applications. Light comes from many sources, although the most common sources are thermal. An incandescent tungsten bulb, an example of a thermal source, emits less than 10% of its energy in the VIS (400–700 nm) and the remainder at IR wavelengths. Other light sources include gas discharge lamps such as neon and mercury-vapor lamps, which emit light at discrete wavelengths characteristic of the atoms in the lamps, light-emitting diodes, and lasers that emit light at specific wavelengths.

8.2 THERMAL SOURCES

A thermal source produces electromagnetic radiation due to the thermal motion of charged particles in the source. Planck's radiation law gives the intensity of the radiation in wavelength interval $\Delta\lambda$ emitted by a blackbody source as a function of wavelength λ and temperature T.

$$I(\lambda,T) = \frac{2\pi hc^2}{\lambda^5} \left(\frac{1}{e^{hc/\lambda k_B T} - 1} \right) \Delta\lambda \tag{8.1}$$

where:
  h is the Planck's constant equal to 6.63×10^{-34} Js
  c is the speed of light in vacuum equal to 3.0×10^8 m/s
  k_B is the Boltzmann constant equal to 1.38×10^{-23} J/K

Equation 8.1 may be expressed in the form

$$I(\lambda,T) = \frac{2\pi h v^3}{c^2} \left(\frac{1}{e^{hv/k_B T} - 1} \right) \Delta v \tag{8.2}$$

where:
  v is the frequency
  Δv is the frequency bandwidth of the light

Wien's displacement law gives the wavelength for maximum intensity

$$\lambda_{max} T = \text{constant } t = 2898 \ \mu mK \tag{8.3}$$

For a blackbody source at 3000 K, λ_{max} is equal to 0.966 μm. Integrating Equation 8.1 with respect to λ or Equation 8.2 with respect to ν, the total intensity I emitted by a blackbody at temperature T is given by

$$I = \left(\frac{2\pi^5 k_B^4}{15 c^2 h^3} \right) T^4 \tag{8.4}$$

The Stefan-Boltzmann radiation law gives the intensity (power per unit area) radiated by a thermal source at temperature T

$$I = \varepsilon \sigma T^4 \tag{8.5}$$

where:
ε is the emissivity of the source
σ is the Stefan-Boltzmann constant equal to 5.67×10^{-8} W/m². K⁴

The value of ε is equal to 1 for a blackbody source. In general, ε depends upon the wavelength of light.

Comparing Equations 8.4 and 8.5

$$\sigma = \left(\frac{2\pi^5 k_B^4}{15 c^2 h^3} \right) \tag{8.6}$$

Figure 8.1 shows the intensity spectrum for a blackbody at 3000 K and $\Delta\lambda$ of 10 nm. The intensity in the VIS region (400–700 nm) in the spectrum of Figure 8.1 is 38.3 W/cm² compared with the total intensity of 459.3 W/cm². Therefore, the VIS efficiency of a blackbody source at 3000 K is only 8.3%.

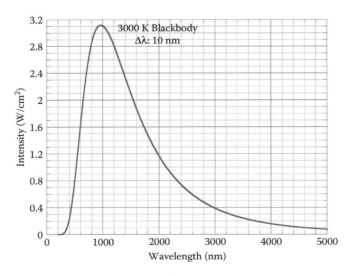

FIGURE 8.1 Spectrum of a blackbody at 3000 K.

TABLE 8.1
Color of a Blackbody Radiator

T (°C)	T (K)	Color
480	753	Faint red glow
580	853	Dark red
730	1003	Bright red, slightly orange
930	1203	Bright orange
1100	1373	Pale yellowish orange
1300	1573	Yellowish white
>1400	>1673	White

Source: Wikipedia, www.wikipedia.org, 2017.

Table 8.1 lists the color of a blackbody radiator at several temperatures (Wikipedia 2017).

The value of ε for the tungsten filament in an incandescent light bulb (3000 K) is 0.35. Therefore, the intensity of the tungsten filament is equal to 161 W/cm². The VIS efficiency of the incandescent tungsten light bulb is about 8%. The remaining radiation (92%) resides in the IR. The glass housing of the light bulb transmits part of the IR radiation and absorbs the rest.

A tungsten filament is used for VIS and IR spectroscopy. A Nernst lamp uses a ceramic rod that is heated to incandescence. The ceramic rod does not need to be enclosed within a vacuum or noble gas environment, because the rod would not further oxidize when exposed to air. Walther Nernst, a German physicist and chemist, developed the Nernst lamp in 1897 at Goettingen University. Nernst lamps are no longer used for IR spectroscopy. At 1250°K–1900°K, Globar (a silicon carbide rod) emits radiation in the 4–15 μm region and is used as a thermal light source for IR spectroscopy. The spectral behavior of a Globar is approximately that of a blackbody.

8.3 GAS-DISCHARGE LAMPS

Gas-discharge lamps generate light by an electrical discharge through an ionized gas. These lamps use noble gases such as argon (Ar), krypton (Kr), neon (Ne), and xenon (Xe). Some gas-discharge lamps include other substances such as mercury (Hg) and sodium (Na), which are vaporized during startup to become a part of the gas mixture. These lamps emit spectral lines characteristic of the gas atoms. Table 8.2 lists the wavelengths (λ) of some of the strongest lines of Hg, Na, Ar, Kr, Ne, and Xe gas-discharge lamps in the UV, VIS, and NIR (NIST 2017).

These gas-discharge lamps are used for the wavelength calibration of spectrometers and are available from several vendors. A deuterium gas-discharge lamp is the preferred light source for UV because of its relatively low intensity in VIS and IR.

TABLE 8.2

Strongest Lines of Hg, Na, Ar, Kr, Ne, and Xe Gas-Discharge Lamps

Hg λ (nm)	Na λ (nm)	Ar λ (nm)	Kr λ (nm)	Ne λ (nm)	Xe λ (nm)
226.2	353.3	750.4	435.6	332.4	488.4
253.7	363.1	763.5	465.9	337.8	529.2
284.8	371.1	794.8	473.9	376.6	533.9
365.0	439.3	800.62	587.1	377.7	541.9
398.4	445.5	801.5	810.4	693.0	597.7
404.7	589.0	810.4	811.3	703.2	605.1
435.8	589.6	811.5	819.0	724.5	699.1
546.1		842.5	826.3	849.5	823.2
		912.3	829.8	865.4	828.0
		965.8	877.7	878.1	881.9

Source: NIST, www.nist.gov/PhysRefData, 2017.

The deuterium lamp in the 200–400 nm UV has higher intensity than that of the hydrogen lamp.

8.4 LIGHT-EMITTING DIODES

A light-emitting diode (LED) is a semiconductor p-n junction diode that emits light due to electroluminescence. Majority carriers on the n-side are electrons, and on the p-side, are holes. Diffusion drives majority carriers on each side to the opposite side of the junction where they are annihilated by the opposite charge carriers. As a result, a space charge region is developed around the junction, resulting in a built-in potential. The resultant electric field impedes majority carrier diffusion to the other side of the junction. An external forward bias reduces the height of the internal barrier, allowing majority carries on each side to be injected to the other side. A region is created around the junction with significant overlap of injected holes and electrons leading to significant radiative recombination. The voltage source replenishes the lost electron-holes pairs by taking electrons from the p-side and supplying electrons to the n-side. This leads to a forward current that increases exponentially with applied voltage. A reverse bias leads to increasing the barrier height, and therefore no current flows in this case. The color of the light is determined by the energy band gap of the semiconductor.

James R. Baird and Garry Pittman developed the infrared LED in 1961 at Texas Instruments, Dallas, Texas (US Patent 3293513). Nick Holonyak Jr. developed the first red LED in 1962 at General Electric (GE), Syracuse, New York using gallium arsenide phosphide (GaAsP) (US Patent 3249473) on a gallium

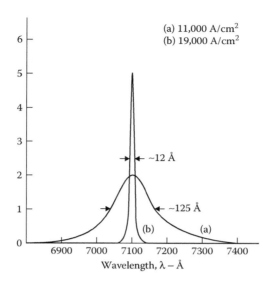

FIGURE 8.2 Spectrum of red LED. (From Holonyak, N. Jr. and Bevacqua, S.F., *Appl. Phys. Lett.,* 1, 82, 1962. With permission.)

arsenide (GaAs) substrate. M. George Craford developed the first yellow LED in 1972 at Monsanto, St. Louis, Missouri using GaAsP (US Patent 3873382). Maruska and Pankove developed the violet LED in 1972 at Radio Corporation of America (RCA), New Jersey using Mg-doped gallium nitride (GaN) films (US Patent 3819974). Shuji Nakamura developed the bright blue LED in 1979 at Nichia Corporation, Tokushima, Japan using GaN (US Patent 5578839). Figure 8.2 shows the spectrum of red LED at 710 nm; curve (b) shows line narrowing upon lasing (Holonyak Jr. 1962).

Table 8.3 shows the LED color along with wavelength range and semiconductor material (Wikipedia 2017).

LEDs have many advantages over incandescent light sources, including lower energy consumption, longer lifetime, improved physical robustness, and faster switching. Even though LEDs have been around for over 50 years, the recent development of white LEDs has made them an attractive replacement for other white light sources. There are two ways to make white LEDs. One is to use red, green, and blue LEDs to generate white light by mixing the three primary colors. The other is to use a yellow or orange phosphor to convert the almost monochromatic light of a blue LED into pure or warm white light. A fraction of the blue light undergoes Stokes shift from shorter to longer wavelengths. A common yellow phosphor is cerium-doped yttrium aluminum garnet (Ce^{3+}:YAG).

LEDs are now used for automotive headlamps, advertising, general lighting, traffic signals, and camera flashes. Currently, LEDs for room lighting are somewhat

TABLE 8.3

LED Color along with Wavelength Range and Semiconductor Material

Color	λ (nm)	Semiconductor
Red	$610 < \lambda < 760$	Gallium arsenide (GaAs)
		Aluminum gallium arsenide (AlGaAs)
Orange	$590 < \lambda < 610$	Gallium arsenide phosphide (GaAsP)
		Aluminum gallium indium phosphide (AlGaInP)
		Gallium phosphide (GaP)
Yellow	$570 < \lambda < 590$	Gallium arsenide phosphide (GaAsP)
		Aluminum gallium indium phosphide (AlGaInP)
		Gallium phosphide (GaP)
Green	$500 < \lambda < 570$	Gallium phosphide (GaP)
		Aluminum gallium indium phosphide (AlGaInP)
		Aluminum gallium phosphide (AlGaP)
		Indium gallium nitride (InGaN)/Gallium nitride (GaN)
Blue	$450 < \lambda < 500$	Zinc Selenide (ZnSe)
		Indium gallium nitride (InGaN)
Violet	$400 < \lambda < 450$	Indium gallium nitride (InGaN)
Ultraviolet (UV)	$\lambda < 400$	Indium gallium nitride (InGaN)
		Aluminum gallium nitride (AlGaN)
		Aluminum gallium indium nitride (AlGaInN)

Source: Wikipedia, www.wikipedia.org, 2017.

TABLE 8.4

Cost of LED, CFL, and ITL

	LED	CFL	ITL
Watts/Bulb	10	14	60
Cost/Bulb (Estimated)	$19	$7	$1.25
Lifespan (hours)	50,000	10,000	1,200
Bulbs needed for 50,000 hours	1	5	42
50,000 hours bulb cost	$19	$35	$53
Cost of electricity (@.10/KWh)	$50	$70	$300
Total cost for 50,000 Hours	$69	$105	$353

Source: Sustainable Supply, www.sustainablesupply.com, 2016.

more expensive than fluorescent lamps. However, the 2016 cost of $19 of an LED for its lifespan of 50,000 hours is less than that of $35 for a compact fluorescent lamp (CFL) and $53 for an incandescent tungsten lamp (ITL), as shown in Table 8.4 (Sustainable Supply 2016). If we include the cost of electricity, LED is even more economical than ITL.

8.5 LASERS

The word *laser* is an acronym for *light amplification* by *stimulated emission* of *radiation*. In 1960, T. H. Maiman at the Hughes Research Laboratories discovered the first laser at 694.3 nm (Maiman 1960). The discovery of the laser has led to the development of whole new industries. The lasers are now found not only in research laboratories but also in many production plants and even on construction sites.

In 1916, Einstein showed that the process of stimulated emission must exist. Positive population inversion must exist between the upper and lower laser levels for achieving laser action. The population inversion is clamped at its threshold value for pump power greater than the threshold pump power. The laser output power is proportional to the pump power minus the threshold pump power. The laser output power is given by

$$P_L = \frac{\ln(R_1 R_2)^{-1/2}}{\alpha L + \ln(R_1 R_2)^{-1/2}} \left(\frac{\tau_{nr}}{\tau_{sp} + \tau_{nr}} \right) (P_P - P_{th}) \tag{8.7}$$

where:
 R_1 and R_2 are reflectance of the mirrors of the laser cavity, respectively
 α and L are the absorption coefficient and length of the laser medium, respectively
 τ_{nr} and τ_{sp} are the nonradiative and spontaneous lifetimes of the laser medium
 P_P and P_{th} are pump power and threshold pump power, respectively

Figure 8.3 shows the energy-level diagram of ruby (Al_2O_3:Cr^{3+}), which was the first laser at 694.3 nm due to the R_1 transition from ^2E to $^4A_2$ (Maiman 1960).

There are several classes of lasers: diode lasers, gas lasers, optically pumped solid-state lasers, dye lasers, chemical lasers, and metal-vapor lasers.

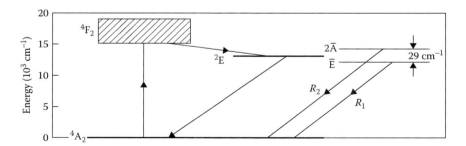

FIGURE 8.3 Energy-level diagram of ruby. (From Maiman, T.H., *Nature*, 187, 493, 1960. With permission.)

TABLE 8.5
Wavelengths of Commercially
Available Diode Lasers

λ (nm)

266
375
405
514, 532
633, 635, 640, 655, 658, 660, 670, 680, 690
705, 780, 785
808, 830, 850, 870
905, 940, 980
1064
1310
1550

8.5.1 DIODE LASERS

When the current in an LED exceeds a threshold value, the emission spectrum of the LED becomes narrow in spectral width and angular divergence, characteristic of a laser. The narrowing of the spectral width is illustrated by curve (b) in Figure 8.2 for a red LED. Diode lasers are now available for several wavelengths listed in Table 8.5.

8.5.2 QUANTUM WELL LASERS

A semiconductor quantum well (QW) is a thin layer (1–20 nm wide) of semiconductor sandwiched (Well/W) between two layers of wider bandgap semiconductors (Barrier/B) in BWB formation (B = 10–20 nm wide). A multiple quantum well (MQW) is a repetition of such a structure. In a Type-I QW, injected electrons and holes each migrate from the B layers to the well, thus increasing the carrier density in that small confined region. Because the radiation recombination is proportional to the square of the electron-hole density, a two-fold increase in the density results in a four-fold increase in the radiative recombination rate. Quantum confinement of electrons and holes also reduces the density of states, so less carrier injection is required to achieve lasing. If a small amount of strain is present in the wells due to lattice mismatch with the barriers, then the density of states of the hole will decrease, thus lowering the threshold current further. A large amount of lattice mismatch, however, is counterproductive since it leads to dislocations and non-radiative recombination. The above discussion applies to interband lasers. In contrast, quantum cascade lasers operate by optical transitions within the conduction band, which leads to lasing in the mid IR region.

8.5.3 GAS LASERS

After the discovery of the laser action in ruby in 1960, a helium-neon (He-Ne) laser, operating at 1.15 μm, was the first cw laser as well as the first gas laser developed by

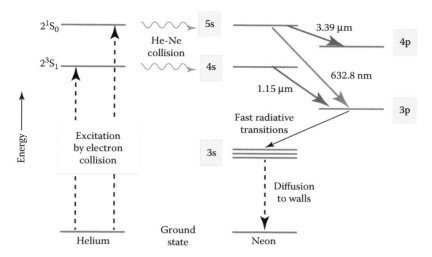

FIGURE 8.4 Energy-level diagram of a He-Ne laser. (Courtesy of Wikipedia, www. wikipedia.org, 2017.)

Javan and coworkers at Bell Labs in 1961 (Javan 1961). The most common type of He-Ne laser operates at 632.8 nm in the visible. The laser transitions occur between the energy levels of neon (Ne). Helium (He) is used to pump neon and obtain a population inversion. The three strongest transitions occur at 0.6328, 1.15, and 3.39 μm.

He atoms are excited to metastable levels 2^1S_0 and 2^3S_1 by direct electron impact. The Ne atom has six more electrons than the He atom. Two of the higher energy states 4s and 5s of Ne have almost the same energy as two of the metastable states of He. With the energy match so close, a collision between a He atom and a neon atom can result in the efficient transfer of energy from the metastable He atom to the unexcited Ne atom. He atoms return to the ground state upon the excitation of the Ne atoms into their excited states. A collision that results in this type of energy transfer is called a resonant collision. The 632.8 nm, 1.15 μm, and 3.39 μm laser lines are due to the 5s to 3p, 4s to 3p, and 5s to 4p transitions, respectively, as shown in Figure 8.4 (Wikipedia 2017).

The rare or noble gases such as argon, xenon, and krypton are all used in gas lasers. Laser transitions in these gases take place between excited states of their ions. Transitions between highly excited states of the singly ionized argon atom can be used to obtain lasing at a number of wavelengths in the visible. The typical ion-laser plasma consists of a high-current density (100–2000 A/cm^2) in the presence of an axial magnetic field (~0.1 T).

William Bridges invented the argon-ion laser in 1964 at Hughes Aircraft (Bridges 1964). Argon-ion lasers emit at 13 wavelengths in the UV, VIS, and near-IR. The strongest laser lines occur at 488.0 and 514.5 nm. Other transitions occur at 351.1, 363.8, 454.5, 457.9, 465.8, 472.7, 476.5, 496.5, 501.7, 528.7, and 1092.3 nm. The argon-ion laser tends to lase simultaneously at several strong lines. A single line operation is achieved by the insertion of a wavelength-dispersion element such as a grating or a prism into the laser cavity.

A krypton-ion laser emits at 13 wavelengths in the visible and near-IR. The laser lines occur at 406.7, 413.1, 415.4, 468.0, 476.2, 482.5, 520.8, 530.9, 568.2, 647.1, 676.4, 752.5, and 799.3 nm. The strong lines are at 530.9, 568.2, 647.1, 752.5, and 799.3 nm.

C. K. N. Patel invented the carbon dioxide (CO_2) molecular gas laser in 1964 at Bell Labs (Patel 1964). The following year he greatly improved the efficiency of the laser by adding nitrogen, which performed the same role as helium did in the He-Ne laser. The CO_2 molecules have energy levels that correspond to the vibration and rotation of the atoms in the molecule. The rotational and vibrational energy levels are much lower than the electronic energy levels. In atomic gases, the energy needed to pump to the higher state is significantly greater than the laser energy. Therefore, the efficiency of the atomic gas lasers is relatively low. Molecular levels are closer to the ground state so that a greater fraction of the pumping energy turns into laser output, which results in the relatively high efficiency of the molecular gas lasers. Also, molecular gas lasers operate at longer wavelengths in the IR region. The CO_2 laser operates in the 10-μm region and can be tuned from about 9.2 to 10.8 μm on a number of vibrational-rotational transitions with an intracavity grating. Figure 8.5 shows the typical spectrum of a grating-tuned 40 W cw CO_2 laser at a pressure of about 10 torr (Rosenbluh 1975).

The CO_2 laser levels are about 2 cm^{-1} apart and about 50 MHz wide at a pressure of 10 torr. The Doppler broadening at room temperature is about 50 MHz. A 1-m long CO_2 laser produces about 50–100 W of cw power. Pulsed outputs in excess of 10 MW peak power in 100 ns pulses are easily obtained from 1-m long lasers operating at atmospheric pressure with transverse excitation in contrast to longitudinal excitation at low

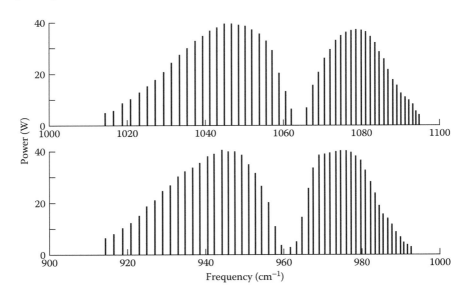

FIGURE 8.5 Spectrum of a grating-tuned 40 W cw CO_2 laser. (From Rosenbluh, M., MS Thesis, Design and evaluation of a continuous-wave, step-tunable far infrared source for solid state spectroscopy, Massachusetts Institute of Technology, Cambridge, MA, 1975.)

pressures of about 10 torr for cw operation. CO_2 lasers are also operated at high pressures of about 15 atmospheres where the transition line width is appreciably larger than the Doppler linewidth, which allows almost continuous tuning over the spectral width.

Gas lasers operate at a large number of wavelengths from the UV to the far IR. Table 8.6 lists wavelengths of the strong laser lines of some gas lasers.

TABLE 8.6
Wavelengths of the Strong Laser Lines of Gas Lasers

Gas	Atom, Ion, or Molecule	Strong Laser Lines (μm)	First Reference
He-Ne	Ne	0.6328, 1.15, 3.39	Javan et al. (1961)
He	He	2.060	Patel et al. (1962)
Ne	Ne	2.102	Patel et al. (1962)
Ne	Ne+	0.2678, 0.2777, 0.3324, 0.3327, 0.3378, 0.3392	Bridges and Chester (1965)
A	A	1.618, 1.694, 1.793, 2.062	Patel et al. (1962)
A	A+	0.3511, 0.4579, 0.4764, 0.4879, 0.5145	Bridges and Chester (1965)
Kr	Kr	1.690, 1.694, 1.784, 1.819, 1.921, 2.116, 2.189	Patel et al. (1962)
Kr	Kr+	0.4619, 0.4762, 0.4766, 0.5208, 0.5682, 0.6471	Bridges (1964)
Xe-He	Xe	2.0261	Patel et al. (1962)
Xe	Xe+	0.4603, 0.5419, 0.5971	Bridges (1964)
Ne-O, Ar-O	O	0.8446	Bennett et al. (1962)
O_2	O+, O++	0.4414, 0.5592	McFarlane (1964)
N_2	N_2	0.8691, 0.8698, 0.8704, 0.8879, 0.8887, 0.8893, 1.0480, 1.2312	Mathias and Parker (1963)
N_2	N+	0.5679	McFarlane (1964)
He-Cl_2	Cl	1.9754, 2.0200	Paananen et al. (1963)
Cl_2	Cl+	0.5218, 0.5221, 0.5302	McFarlane (1964b)
Ar-Br_2	Br	0.8446	Patel et al. (1964)
Br_2	Br+	0.5182, 0.5332	Keefe and Graham (1965)
I_2	I	3.236, 3.431	Rigden and White (1963)
He-I_2	I+	0.5407, 0.5678, 0.5761, 0.6127, 0.6585, 0.7032	Fowles and Jensen (1964)
SF_6	S	1.0455, 1.0628	Patel et al. (1964)
He-S	S+	0.53207, 0.53457, 0.54328, 0.54538, 0.56400	Fowles et al. (1965)
He-P	P+	0.602418, 0.604312, 0.784563	Fowles et al. (1965)
He-Zn	Zn+	0.492404, 0.775786	Fowles et al. (1965)
He-Cd	Cd+	0.533740, 0.537804	Fowles et al. (1965)
Hg	Hg	1.529, 1.813	Rigden and White (1963)
He-Hg	Hg+	0.5678, 0.6150	Bell (1964)
CO_2-N_2	CO_2	9.2–10.8	Patel (1964)
CO	CO	5.2–6.0	Patel and Kerl (1964)

8.5.4   OPTICALLY PUMPED SOLID-STATE LASERS

A solid-state laser consists of a glass or crystalline host material doped with impurities such as chromium, neodymium, titanium, or ytterbium. Solid-state lasers are optically pumped using either a flash lamp, an arc lamp, or a laser diode. Diode-pumped solid-state (DPSS) lasers are more efficient. Also, DPSS lasers have a compact setup, long lifetime, and very good beam quality. Therefore, DPSS lasers have become more popular. DPSS lasers are either end-pumped or side-pumped, as shown in Figures 8.6 and 8.7, respectively (RP Photonics).

Demonstrated in 1960 by Maiman at Hughes Research Laboratories, the flash-lamp-pumped chromium-doped ruby laser was the very first laser. Besides ruby, a number of other crystals doped with impurity atoms have been shown to lase. Table 8.7 lists host material, laser ion, and wavelengths of some optically pumped solid-state lasers.

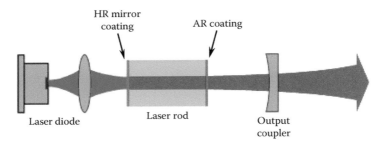

FIGURE 8.6   Laser diode end-pumped solid-state laser, converting pump light (shown as blue) into laser light (shown as red). (Courtesy of RP photonics, www.rp-photonics.com, 2017.)

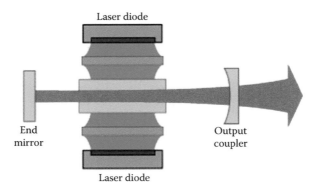

FIGURE 8.7   Laser diode side-pumped solid-state laser, converting pump light (shown as blue) into laser light (shown as red). (Courtesy of RP photonics, www.rp-photonics.com.)

TABLE 8.7

Host Material, Laser Ion, and Wavelengths of Optically Pumped Solid-State Lasers

Host Material	Laser Ion	λ (μm)	First Reference
Al_2O_3	Cr^{3+}	0.6493	Maiman (1960)
CaF_2	Er^{3+}	1.6	Pollack (1963)
$CaWO_4$	Er^{3+}	1.612	Kiss and Duncan Jr. (1962)
LaF_3	Er^{3+}	1.611	Krupke and Gruber (1964)
Y_2O_3	Eu^{3+}	0.6113	Wickersheim and Lefever (1964)
Glass	Gd^{3+}	0.3125	Gandy and Ginther (1962)
CaF_2	Ho^{3+}	2.09	Sabisky and Lewis (1963)
$CaWO_4$	Ho^{3+}	2.05	Johnson et al. (1962a)
BaF_2	Nd^{3+}	1.06	Payne et al. (1991)
SrF_2	Nd^{3+}	1.043	Payne et al. (1991)
CaF_2	Nd^{3+}	1.06	Payne et al. (1991)
$CaWO_4$	Nd^{3+}	1.058	Johnson et al. (1962b)
$CAWO_4$	Nd^{3+}	0.90, 1.35	Johnson and Thomas (1963)
LaF_3	Nd^{3+}	0.172	Waynant and Klein (1985)
YAG	Nd^{3+}	1.053–1.079	Geusic et al. (1964)
$CaWO_4$	Pr^{3+}	1.047	Yariv et al. (1962)
LaF_3	Pr^{3+}	0.5985	Solomon and Mueller (1963)
CaF_2	Sm^{2+}	0.7082	Sorokin and Stevenson (1961)
SrF_2	Sm^{2+}	0.6969	Sorokin et al. (1962)
Al_2O_3	Ti^{3+}	0.662–0.950	Moulton (1986)
CaF_2	Tm^{2+}		Kiss and Duncan Jr. (1962)
$CaWO_4$	Tm^{3+}	1.911	Johnson et al. (1962a)
CaF_2	U^{3+}	2.49	Sorokin and Stevenson (1960)
CaF_2	U^{3+}	2.51–2.61	Wittke et al. (1963)

8.5.5 DYE LASERS

The active medium of a dye laser consists of an organic dye dissolved in a solvent. When the dye is optically pumped with a flash lamp or a laser, it emits radiation at longer wavelengths. Both the absorption and emission bands are usually broad, as shown in Figure 8.8, for rhodamine 6G (Wikipedia 2017).

The emissions of dye lasers are inherently broad. Narrow linewidth dye lasers are obtained using gratings, prisms, and etalons as the wavelength-tuning element in the laser cavity. Some of the laser dyes are rhodamine 6G, fluorescein, and coumarin, which are tunable over the range 540–680 nm, 530–560 nm, and 490–620 nm, respectively.

The small-signal gain of most dye lasers is extremely high. Therefore, only a small amount of active medium is needed. However, the intense absorption and subsequent heating of the small volume of the dye necessitates a continuous and rapid change of the pumped volume. A liquid jet design of a ring dye laser is shown in

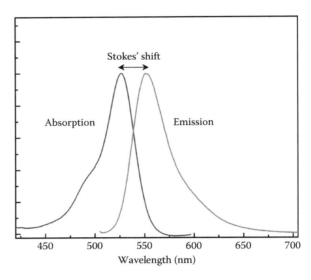

FIGURE 8.8 Absorption and emission bands of rhodamine 6G. (Courtesy of Wikipedia, www.wikipedia.org, 2017.)

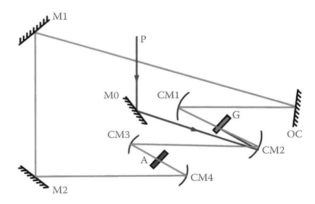

FIGURE 8.9 Schematic of a ring dye laser. (Courtesy of Wikipedia, www.wikipedia.org, 2017.)

Figure 8.9 (Wikipedia 2017). P is the pump laser. G is the gain dye jet. M0, M1, and M2 are planar mirrors. OC is the output coupler. CM1, CM2, CM3, and CM4 are curved mirrors.

8.5.6 CHEMICAL AND METAL-VAPOR LASERS

A chemical laser obtains its energy from a chemical reaction. Common examples of chemical lasers are (1) chemical oxygen iodine laser (COIL), (2) hydrogen fluoride laser (2.7–2.9 μm), (3) deuterium fluoride laser (3-6–4.2 μm), and (4) oxygen-iodine laser (1.315 μm). The cw HF laser was first demonstrated in 1969 (Spencer 1969).

Metal vapor lasers include helium-cadmium laser (325 nm, 441.6 nm), helium-mercury laser (567 nm, 615 nm), helium-silver laser (224.3 nm), strontium vapor laser (430.5 nm), neon-copper laser (248.6 nm), copper vapor laser (510.6 nm, 578.2 nm), gold vapor laser (627 nm), and manganese vapor laser (534.1 nm).

8.5.7 Fiber Lasers

The gain medium in a fiber laser is an optical fiber doped with rare-earth elements such as erbium, ytterbium, and neodymium. The gain medium is the core of the fiber, surrounded by two layers of cladding. The multimode pump beam propagates in the inner cladding layer while the outer cladding layer keeps this pump light confined. Bragg gratings are used to replace conventional dielectric mirrors for optical feedback. Fiber lasers are very useful because of (1) vibrational stability, (2) high beam quality, (3) high efficiency (70%–80%), and (4) high cw power (Motes 2009). Heat is distributed over the long length of the fiber; this protects the fiber from getting hot.

8.6 HOMEWORK PROBLEMS

8.1 Deduce Equation 8.3 for Wien's displacement law.

8.2 A blackbody filament operates at a temperature of 1500 K. Calculate wavelength for the peak intensity of the radiation from this filament. Also, calculate the intensity/nm at this wavelength.

8.3 The light output of a 10 W LED bulb is equivalent to that of a 60 W tungsten bulb. Determine the efficiency of the LED bulb assuming that the efficiency of the tungsten lamp in the visible is 8%.

8.4 A 60-W incandescent bulb runs 10 h/day. The lifetime and cost of this incandescent bulb are 3500 h and $1.00, respectively. The cost of electricity is assumed to be 20 cents/kWh. Calculate the total cost of operation, which includes replacement of bulbs, for a period of 10 years.

8.5 A 9-W LED bulb with visible output equivalent to that of a 60-W incandescent bulb runs 10 h/day. The lifetime and cost of the LED bulb are >10 years and $10.00, respectively. The cost of electricity is assumed to be 20 cents/kWh. Calculate the total cost of operation, which includes the cost of the LED bulb, for a period of 10 years.

8.6 Determine the average separation (cm^{-1}) between the CO_2 laser lines in the 9.6 and 10.6 μm bands using the CO_2 laser spectrum shown in Figure 8.5.

8.7 Determine the laser output power using values of 1.00 for R_1, 0.95 for R_2, 0.02 for αL, 0.80 for $\tau_{nr}/(\tau_{sp} + \tau_{nr})$, 10.0 W for P_P, and 1.0 W for P_{th}.

9 Light Detectors

9.1 INTRODUCTION

Light detectors are used for the detection and power measurement of light. There are two types of optical detectors. Thermal detectors make use of the heating effect of light. As such their response is determined only by the absorbed power, which is largely independent of the wavelength of light. Photon detectors, on the other hand, are strongly dependent upon the wavelength of light. Figure 9.1 shows comparison of the signals of idealized thermal and photon detectors.

9.2 THERMAL DETECTORS

The operation of thermal detectors depends on the heating produced by light in an absorbing surface of the detector. The resultant increase in the temperature of the detector may be measured by the change of its resistance, as in a bolometer, or by the change of thermoelectric potential at a junction between two metals, as in a thermocouple. A large temperature change for a given power of incident light requires a detector element, which has a small thermal capacity and is well insulated from its surroundings.

9.2.1 BOLOMETERS

Bolometers are small infrared sensors, which are thermal detectors. Samuel Pierpont Langley, an American astronomer, invented the bolometer in 1878. A bolometer consists of an absorptive element, which is connected to a thermal reservoir through a thermal link. Light incident upon the absorptive element raises its temperature above that of the reservoir. Most bolometers make use of the thermistor material, which is composed of the oxides of manganese, cobalt, and nickel. The resistance change of the bolometer is defined by

$$\alpha = \frac{1}{\rho}\frac{d\rho}{dT} \qquad (9.1)$$

where:
ρ is the resistivity
T is the temperature of the bolometer

The value of α is around 0.01/K at room temperature. The thermal time constant τ of the bolometer is around a few milliseconds. Figure 9.2 shows the bolometer circuit for the measurement of the signal voltage, when the light is chopped with the on/off time $t \gg \tau$.

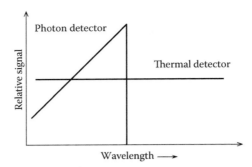

FIGURE 9.1 Idealized thermal and photon detectors.

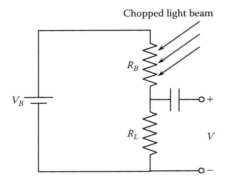

FIGURE 9.2 Circuit for the measurement of the bolometer signal voltage.

For $R_L = R_B$, the signal voltage is given by

$$V_s = -\frac{1}{4}V_b\Delta T \tag{9.2}$$

where ΔT is the temperature increase during the on cycle of the light. The noise equivalent power (NEP) of the room-temperature bolometer is $\sim 1 \times 10^{-8}$ W/Hz$^{1/2}$.

A microbolometer is a type of bolometer that is used as a detector in a thermal camera. Vanadium oxide (V_2O_3) microbolometer arrays are available in the following sizes: 160×120, 320×240, 640×480, and 1024×768 pixels. IR radiation changes the resistance of V_2O_3. These changes in resistance are measured and processed into temperature changes.

Carbon and germanium bolometers, which are cooled to liquid helium temperatures, are used for spectroscopic applications in the infrared. Boyle and Rodgers, Jr. developed the carbon bolometer in 1959 (Boyle and Rodgers 1959). Low developed the germanium bolometer in 1961 (Low 1961). The NEP of the carbon (2.1 K, 0.2 cm²) and germanium (2.15 K, 0.15 cm²) bolometers was measured to be 1×10^{-11} and 5×10^{-13} W/Hz$^{1/2}$, respectively. Specific detectivity is given by

$$D^* = \frac{\sqrt{A_d}}{NEP} \tag{9.3}$$

where A_d is the area of the detector. Therefore, D* of the carbon and germanium bolometers is equal to 4.5×10^{10} and 7.7×10^{11} cmHz$^{1/2}$/W. By decreasing the thickness of the bolometer element, Corsi et al. have reported obtaining a lower NEP of less than 1×10^{-12} W/Hz$^{1/2}$ for the carbon bolometer (Corsi 1973).

9.2.2 THERMOCOUPLES

A thermocouple is a thermal detector, which is used to measure the light power incident upon it. It consists of two dissimilar metals that form two electrical junctions. By blackening one of the two junctions, incident light is absorbed resulting in the temperature rise of this junction. This temperature increase produces a voltage because of the thermoelectric effect. To be useful, a thermocouple must respond rapidly to light. Also, the receiver area of the thermocouple should reasonably match the image of the light source. Properties of a spectrometer-type thermocouple developed by Perkin-Elmer (Liston 1947; Waard and Wormser 1959) are given below:

Receiver Area: 2×0.2 mm²;
DC resistance: 10 ohms;
Responsivity at 13 Hz chopping: 2 V/W;
NEP: 3×10^{-10} W/ Hz$^{1/2}$;
D*: 2×10^8 cmHz$^{1/2}$/W

9.2.3 PYROELECTRIC DETECTORS

Pyroelectric materials, such as barium titanate ($BaTiO_3$) and triglycine sulfate (TGS), generate electricity when heated. Cooper first made analysis of the infrared pyroelectric detectors in 1962 (Cooper 1962). The NEP of the pyroelectric detectors is about 1×10^{-8} W/Hz$^{1/2}$ (Ludlow 1967). This value of NEP is comparable to that of a room-temperature bolometer.

9.3 PHOTON DETECTORS

A photon detector is based upon the detection of photons. Photon detectors are faster and more efficient than thermal detectors. Important classes of photon detectors are:

1. Photoconductive are based on the photoconductive effect in which the incident photons produce free electrons and holes in a semiconductor.
2. Photovoltaic measure the voltage produced by the separation of electrons and holes in a semiconductor p-n junction.
3. Photoemissive are based on the photoelectric effect in which the incident photons release electrons from the surface of the detector material.

9.3.1 Photoconductive Detectors

Photoconductive detectors are most widely used for the detection of infrared light. Photoconductive detectors were among the first solid-state optical detectors. There are two types of photoconductive detectors: intrinsic and extrinsic. In the intrinsic case, the energy of the absorbed photon produces an electron-hole pair. In the extrinsic case, free electrons or free holes are produced. In either case, the maximum wavelength of light that can be detected is given by

$$\lambda_{max} = \frac{1.2406}{E_p} \tag{9.4}$$

where:

λ_{max} is in microns
E_p is the minimum photon energy in electron volts

Figure 9.3 shows a photoconductor of length L in a series with a load resistor connected to a bias battery of voltage V_0 with optical power P incident upon it.

Photocurrent in the presence of optical power P is given by (Kingston 1995)

$$i = \frac{\eta e G_p \lambda P}{hc} \tag{9.5}$$

where:

η is the quantum efficiency, which is a function of the wavelength and temperature of the detector
e is the electron charge
G_p is the photoconductive gain
λ is the wavelength of light
h is the Planck's constant
c is the velocity of light in vacuum

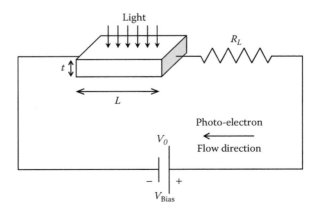

FIGURE 9.3   Photoconductor with optical power P incident upon it.

TABLE 9.1

D* of Semiconductor Detectors Operating in the Photoconductive Mode at Room Temperature

Semiconductor	Peak λ (μm)	D* (cm Hz$^{1/2}$/W)
Si	1.0	6.4×10^{15}
Ge	1.5	2.54×10^{14}
PbS	2.2	4.0×10^{12}
PbSe	3.4	4.4×10^{10}
PbTe	5.9	1.3×10^{12}
InAs	3.6	6.3×10^{12}
InSb	6.6	4.4×10^{11}
Te	3.6	1.8×10^{12}

Source: Kruse, P.W., McGlauchlin, L.D., and McQuistan, R.B., *Elements of Infrared Technology*, John Wiley & Sons, New York, 1962.

If the photoconductor is placed in a series with a load resistor R_L, then the voltage changes across the load resistor in the presence of the optical power P are given by

$$\Delta V_{R_L} = iR_L \tag{9.6}$$

Intrinsic photoconductive detectors consist of a number of semiconductors such as Si, Ge, PbS, PbSe, PbTe, CdS, HgCdTe, GaAs, InGaAs, and so on. Extrinsic photoconductive detectors are made by doping Si or Ge with impurities, such as As or Cu, and are operated at liquid helium temperatures. The extrinsic photoconductive detectors are used at IR wavelengths >10 μm.

Table 9.1 lists computed D* at the assumed peak wavelength of semiconductor detectors operating in the photoconductive mode at room temperature (Kruse 1962).

Quantum well infrared photodetectors (QWIP) are another class of photoconductive detectors that operate in the mid IR. In these devices, photon absorption occurs from a lower energy level to a higher level within the same quantum well in the conduction band. An applied bias will collect the charge from the wells leading to a photocurrent. Two-dimensional arrays of these detector elements are integrated with corresponding read-out integrated circuits (ROIC) and form the imaging element of modern IR imaging cameras. The detector elements need to be cooled to near 80°K. The handheld cameras come with their own integrated cooling system.

9.3.2 PHOTOVOLTAIC DETECTORS

Photodiodes produce a voltage generated by the photo-created carriers in the junction. This voltage is given by (Kingston 1995)

$$V = \frac{k_B T}{e I_S} \quad \text{for } i \ll I_S \tag{9.7}$$

where:
k_B is the Boltzmann constant
T is the temperature of the photodiode
I_S is the saturation current of the photodiode
i is the photocurrent

This voltage can be measured without need for a bias supply or load resistor.
The spectral photodiode responsivity is defined by

$$R_\lambda = \frac{i}{P_\lambda} \qquad (9.8)$$

where P_λ is power of the light of wavelength λ incident upon the photodiode. The value of R_λ for a silicon photodiode is about 0.7 A/W at 850 nm.

9.3.3 AVALANCHE PHOTODIODES

Avalanche photodiodes (APDs) are photodiodes with an internal gain mechanism similar to a photomultiplier tube. A high reverse bias voltage is applied to the photodiodes to create a strong electric field. When an incident photon creates an electron-hole pair, the electric field accelerates the electron and holes producing secondary electrons and holes by impact ionization. The ionization coefficients α_n and α_p for the electrons and holes, respectively, increase rapidly with electric field values of the order of 10^5 V/cm. The avalanche multiplication factor is given by (Kingston 1995)

$$M = \frac{1}{1 - \alpha w} \qquad (9.9)$$

where:
α is equal to α_n for electrons
α_p for holes

In general, the M factor increases with an increase in the reverse bias voltage. Also, the M factor increases with a decrease in the temperature of the photodiode. Temperature-compensated APDs use integrated thermistors that adjust the bias voltage to compensate the effect of temperature changes on the M factor. Si APDs can produce M factors up to several hundred. Si APDs are used for the UV, VIS, and NIR from 200 to 1000 nm. Ge APDs are used for the NIR from 800 to 1500 nm. InGaAs APDs are used for the NIR from 1000 to 1650 nm. GaN APDs are used for the UV. Several APDs and APD arrays are commercially available.

Geiger-mode APDs can detect a single photon with a jitter of a few tens of picoseconds. The name Geiger-mode, as opposed to linear mode in an APD, is used as an analogy to the Geiger counter. The Geiger-mode APDs operate well above the breakdown voltage where a single charge carrier injected into the depletion layer can trigger a self-sustaining avalanche.

9.3.4 Photomultiplier Tubes

The invention of the photomultiplier tube (PMT) is based on the discovery of two separate phenomena: (1) the photoelectric effect, which is the emission of an electron from the surface of a photocathode by a photon; and (2) the secondary emission of electrons by an energetic electron striking an electrode. A PMT is a vacuum photoemissive detector, which contains a photocathode that emits the primary photoelectrons and a number of secondary emitting stages called dynodes. Multiplication of the primary photoelectrons occurs because of the secondary electrons. The multiplication factor is given by

$$M = \delta^N \qquad (9.10)$$

where:
  δ is the number of secondary electrons emitted per primary electron
  N is the number of dynode stages

δ may have a value of about 4. The factor N may have a value of 10. In this case, M has a value of about 1×10^6. The secondary emission process causes a spread in the transit times of the electrons, which limits the frequency response to about 100 MHz.

Figure 9.4 shows the construction of a PMT, which consists of a window for the light to pass through, a photocathode that produces the primary photoelectrons, a focusing electrode that accelerates and focuses the photoelectrons onto the first dynode, and 9 other dynodes. Each of the dynodes is biased approximately 100 V positive with respect to the preceding stage. Table 9.2 lists the common photocathode materials.

Figure 9.5 shows the secondary emission ratio as a function of the accelerating voltage for various dynode materials.

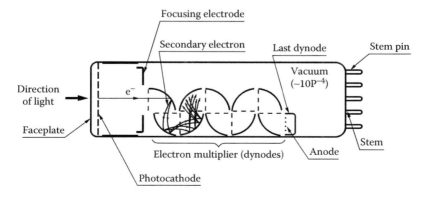

FIGURE 9.4 Construction of a photomultiplier tube. (Courtesy of Hamamatsu Photonics.)

TABLE 9.2
Common Photocathode Materials

Material	λ (nm)
Ag-O-Cs (S1)	300–1200
GaAs:Cs	300–850
InGaAs:Cs	900–1000
Sb-Cs (S11)	200–700
Bialkali (Sb-K-Cs, Sb-Rb-Cs)	200–700
Multialkali (Na-K-Sb-Cs) (S20)	200–850
Solar-Blind (Cs-I)	<200

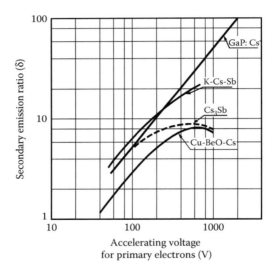

FIGURE 9.5 Secondary emission ratio for several dynode materials. (Courtesy of Hamamatsu Photonics.)

9.3.5 SILICON PHOTOMULTIPLIERS

Silicon photomultipliers (SiPMs) are a replacement for the classical PMTs and APDs (First Sensor 2015). A SiPM consists of an APD array on a common Si substrate. The dimension of each single APD called a micro-cell is 20–100 μm. The density of the micro-cells can be up to 1000 per square mm. Each micro-cell in a SiPM operates in the Geiger-mode. The bias voltage applied to a micro-cell is 20–100 V, which is significantly lower than the 1000–1500 V required for a PMT. Because of their small dimensions, SiPMs are compact, light, and robust. The photon detection efficiency and gain of the SiPM is similar to that of the PMT. The frequency response of a SiPM is about 10 GHz, which is 100x larger than that of a PMT. In contrast to that of a PMT, the SiPM signal output is independent of the external magnetic fields.

9.4   PHOTODETECTOR NOISE

There are three sources of photodetector noise: (1) shot noise, which is proportional to the square root of the optical power P_S incident upon the photodetector; (2) background noise, which is proportional to the square root of the background radiation power P_B incident upon the photodetector; and (3) thermal noise or Johnson noise, which is proportional to the square root of T/R_{in} where T is temperature and R_{in} is the effective input resistance of the photodetector. The signal current i_S and the noise current i_n of the photodetector are given by

$$i_S = \eta e \frac{P_S}{h\nu} \tag{9.11}$$

where:
η is the quantum efficiency of the photodetector
ν is the light frequency.

$$i_n = \left[\left(2\eta e^2 P_S + i_d^2 \right) \Delta f \right]^{1/2} \tag{9.12}$$

where:
$\Delta f = 1/2\tau$ is the frequency bandwidth
τ is the integration time for the measurement

The first term in Equation 9.12 represents the shot noise and the second term represents the dark noise. The signal-to-noise ratio is given by

$$\text{SNR} = \frac{i_S}{i_n} \tag{9.13}$$

9.5   HOMEWORK PROBLEMS

9.1 Determine the NEP of a 1×1 mm^2 germanium bolometer with D* of 8×10^{11} cmHz$^{1/2}$/W for an integration time of 1 s.

9.2 The germanium bolometer in Problem 9.1 is used for the detection of 10-μm light. Determine the number of 10-μm photons incident upon the detector to obtain a signal-to-noise ratio of 10.

9.3 The NEP of a 2×0.2 mm^2 thermocouple is 3×10^{-10} W/Hz$^{1/2}$. Calculate the signal power required to obtain a signal-to-noise ratio of 3 for an integration time of 10 s.

9.4 The NEP of a pyroelectric detector is 1×10^{-8} W/Hz$^{1/2}$. Determine the signal power required to achieve a signal-to-noise ratio of 10 for an integration time of 30 s.

9.5 D* of a PbS detector at 2.2 μm is 4.0×10^{12} cm Hz$^{1/2}$/W. The area of this detector is 2×0.2 mm^2. Determine the NEP of this detector for an integration time of 10 s.

9.6 The spectral responsivity of a silicon photodiode is 0.7 A/W at 800 nm. Calculate the value of the photocurrent i for 1×10^{-13} W of 800 nm light power incident upon the photodiode.

9.7 The avalanche multiplication factor M is 100 for an APD. Calculate the value of the ionization coefficient α_n for the electrons, assuming that the ionization coefficient α_p for the holes is negligible.

9.8 A PMT contains 10 dynode stages. The number of secondary electrons emitted per primary electron incident upon a dynode is 5. Calculate the multiplication factor M of the PMT.

9.9 The quantum efficiency η of a photodetector is 70%. Calculate the value of the signal current i_s for 1 pW signal power of 600 nm.

9.10 The quantum efficiency η of a photodetector is 70%. Calculate the value of the shot noise current for signal power of 1 pW and integration time of 10 s.

Appendix

SOLUTIONS FOR HOMEWORK PROBLEMS

1.1 a. Focal length f is given by

$$\frac{1}{f} = (n-1)\left[\frac{1}{R_1} - \frac{1}{R_2} + \frac{(n-1)t_C}{nR_1R_2}\right] \qquad \text{(A1.1a)}$$

Substituting values of 1.5 for n, 5.0 cm for R_1, −10.0 cm for R_2, and 1.0 cm for t_C in Equation A1.1a, we obtain a value of 6.818 cm for f.

b. δ_1 is given by

$$\delta_1 = -ft_C\left(\frac{n-1}{nR_2}\right) \qquad \text{(A1.1b)}$$

Substituting values of 6.818 cm for f, 1.0 cm for t_C, 1.5 for n, and −10 cm for R_2 in Equation A1.1b, we obtain a value of 0.227 cm for δ_1.

c. δ_2 is given by

$$\delta_2 = -ft_C\left(\frac{n-1}{nR_1}\right) \qquad \text{(A1.1c)}$$

Substituting values of 6.818 cm for f, 1.0 cm for t_C, 1.5 for n, and 5 cm for R_1 in Equation A1.1c, we obtain a value of −0.454 cm for δ_2.

d. Front focal length f_F is given by

$$f_F = f - \delta_1 \qquad \text{(A1.1d)}$$

Substituting values of 6.818 cm for f and 0.227 cm for δ_1 in Equation A1.1d, we obtain a value of 6.591 cm for f_F.

e. Back focal length f_B is given by

$$f_B = f + \delta_2 \qquad \text{(A1.1e)}$$

Substituting values of 6.818 cm for f and −0.454 cm for δ_2 in Equation A1.1e, we obtain a value of 6.364 cm for f_B.

f. Edge thickness t_E is given by

$$t_E \approx t_C - \frac{D^2}{8}\left(\frac{1}{R_1} - \frac{1}{R_2}\right) \qquad \text{(A1.1f)}$$

127

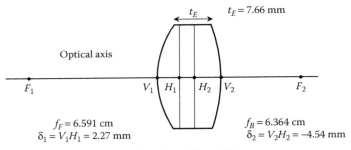

$$f_F = 6.591 \text{ cm}$$
$$\delta_1 = V_1 H_1 = 2.27 \text{ mm}$$

$$f_B = 6.364 \text{ cm}$$
$$\delta_2 = V_2 H_2 = -4.54 \text{ mm}$$

$$f = F_1 H_1 = H_2 F_1 = 6.818 \text{ mm}$$

FIGURE A1.1   Bioconvex lens surfaces, principal planes, and points V_{11}, V_{22}, H_{11}, H_{22}, F_{11}, and F_{22}.

where D is the diameter of the lens. Substituting values of 10 mm for t_C, 25 mm for D, 50 mm for R_1, and -100 mm for R_2 in Equation A1.1f, we obtain a value of 7.66 mm for t_E.

The lens surfaces, optical axis, principal planes, and points V_1, V_2, H_1, H_2, F_1, and F_2 are shown in Figure A1.1.

1.2 a. Focal length f is given by

$$\frac{1}{f} = (n-1)\left[\frac{1}{R_1} - \frac{1}{R_2} + \frac{(n-1)t_C}{nR_1R_2} \right] \tag{A1.2a}$$

Substituting values of 1.5 for n, -5.0 cm for R_1, 10.0 cm for R_2, 1.0 cm for t_C in Equation A1.2a, we obtain a value of -6.522 cm for f.

b. δ_1 is given by

$$\delta_1 = -f t_C \left(\frac{n-1}{nR_2} \right) \tag{A1.2b}$$

Substituting values of -6.522 cm for f, 1.0 cm for t_C, 1.5 for n, and 10 cm for R_2 in Equation A1.2b, we obtain a value of 0.217 cm for δ_1.

c. δ_2 is given by

$$\delta_2 = -f t_c \left(\frac{n-1}{nR_1} \right) \tag{A1.2c}$$

Substituting values of -6.522 cm for f, 1.0 cm for t_C, 1.5 for n, and -5 cm for R_1 in Equation A1.2c, we obtain a value of -0.435 cm for δ_2.

d. Front focal length f_F is given by

$$f_F = f - \delta_1 \tag{A1.2d}$$

Substituting values of -6.522 cm for f and 0.217 cm for δ_1 in Equation A1.2d, we obtain a value of -6.739 cm for f_F.

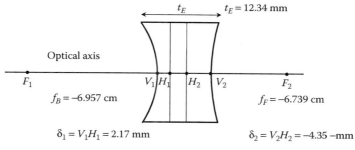

$t_E = 12.34$ mm

Optical axis

F_1 $V_1 | H_1$ $H_2 | V_2$ F_2

$f_B = -6.957$ cm $f_F = -6.739$ cm

$\delta_1 = V_1 H_1 = 2.17$ mm $\delta_2 = V_2 H_2 = -4.35$ -mm

$f = F_1 H_1 = H_2 F_1 = -6.522$ mm

FIGURE A1.2 Bioconcave lens surfaces, principal planes, and points V_{11}, V_{22}, H_{11}, H_{22}, F_{11}, and F_{22}.

e. Back focal length f_B is given by

$$f_B = f + \delta_2 \tag{A1.2e}$$

Substituting values of -6.522 cm for f and -0.435 cm for δ_2 in Equation A1.2e, we obtain a value of -6.957 cm for f_B.

f. Edge thickness t_E is given by

$$t_E \approx t_C - \frac{D^2}{8}\left(\frac{1}{R_1} - \frac{1}{R_2}\right) \tag{A1.2f}$$

where D is the diameter of the lens. Substituting values of 10 mm for t_C, 25 mm for D, -50 mm for R_1, and 100 mm for R_2 in Equation A1.2f, we obtain a value of 12.34 mm for t_E.

The lens surfaces, optical axis, principal planes, and points V_1, V_2, H_1, H_2, F_1, and F_2 are shown in Figure A1.2.

1.3 a. Focal length f is given by

$$\frac{1}{f} = (n-1)\left[\frac{1}{R_1} - \frac{1}{R_2} + \frac{(n-1)t_C}{nR_1R_2}\right] \tag{A1.3a}$$

Substituting values of 1.5 for n, 2.5 cm for R_1, 5.0 cm for R_2, 1.0 cm for t_C in Equation A1.3a, we obtain a value of 8.823 cm for f.

b. δ_1 is given by

$$\delta_1 = -ft_C\left(\frac{n-1}{nR_2}\right) \tag{A1.3b}$$

Substituting values of 8.823 cm for f, 1.0 cm for t_C, 1.5 for n, and 5.0 cm for R_2 in Equation A1.3b, we obtain a value of -0.588 cm for δ_1.

c. δ_2 is given by

$$\delta_2 = -f t_C \left(\frac{n-1}{n R_1} \right) \tag{A1.3c}$$

Substituting values of 8.823 cm for f, 1.0 cm for t_C, 1.5 for n, and 2.5 cm for R_1 in Equation A1.3c, we obtain a value of −1.176 cm for δ_2.

d. Front focal length f_F is given by

$$f_F = f - \delta_1 \tag{A1.3d}$$

Substituting values of 8.823 cm for f and −0.588 cm for δ_1 in Equation A1.3d, we obtain a value of 9.411 cm for f_F.

e. Back focal length f_B is given by

$$f_B = f + \delta_2 \tag{A1.3e}$$

Substituting values of 8.823 cm for f and −1.176 cm for δ_2 in Equation A1.3e, we obtain a value of 7.647 cm for f_B.

f. Edge thickness t_E is given by

$$t_E \approx t_C - \frac{D^2}{8} \left(\frac{1}{R_1} - \frac{1}{R_2} \right) \tag{A1.3f}$$

where D is the diameter of the lens. Substituting values of 10 mm for t_C, 25 mm for D, 25 mm for R_1, and 50 mm for R_2 in Equation A1.3f, we obtain a value of 8.43 mm for t_E.

The lens surfaces, optical axis, principal planes, and points V_1, V_2, H_1, H_2, F_1, and F_2 are shown in Figure A1.3.

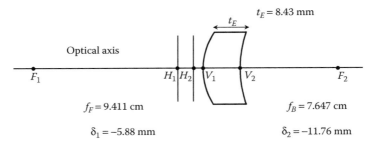

FIGURE A1.3 Positive meniscus lens surfaces, principal planes, and points V_{11}, V_{22}, H_{11}, H_{22}, F_{11}, and F_{22}.

1.4 a. Focal length f is given by

$$\frac{1}{f} = (n-1)\left[\frac{1}{R_1} - \frac{1}{R_2} + \frac{(n-1)t_C}{nR_1R_2}\right] \qquad \text{(A1.4a)}$$

Substituting values of 1.5 for n, 5.0 cm for R_1, 2.5 cm for R_2 1.0 cm for t_C in Equation A1.4a, we obtain a value of -11.538 cm for f.

b. δ_1 is given by

$$\delta_1 = -ft_C\left(\frac{n-1}{nR_2}\right) \qquad \text{(A1.4b)}$$

Substituting values of -11.538 cm for f, 1.0 cm for t_C, 1.5 for n, and 2.5 cm for R_2 in Equation A1.4b, we obtain a value of 1.538 cm for δ_1.

c. δ_2 is given by

$$\delta_2 = -ft_C\left(\frac{n-1}{nR_1}\right) \qquad \text{(A1.4c)}$$

Substituting values of -11.538 cm for f, 1.0 cm for t_C, 1.5 for n, and 5.0 cm for R_1 in Equation A1.4c, we obtain a value of 0.769 cm for δ_2.

d. Front focal length f_F is given by

$$f_F = f - \delta_1 \qquad \text{(A1.4d)}$$

Substituting values of -11.538 cm for f and 1.538 cm for δ_1 in Equation A1.4d, we obtain a value of -13.076 cm for f_F.

e. Back focal length f_B is given by

$$f_B = f + \delta_2 \qquad \text{(A1.4e)}$$

Substituting values of -11.538 cm for f and 0.769 cm for δ_2 in Equation A1.4e, we obtain a value of -10.769 cm for f_B.

f. Edge thickness t_E is given by

$$t_E \approx t_C - \frac{D^2}{8}\left(\frac{1}{R_1} - \frac{1}{R_2}\right) \qquad \text{(A1.4f)}$$

where D is the diameter of the lens. Substituting values of 10 mm for t_C, 25 mm for D, 50 mm for R_1, and 25 mm for R_2 in Equation A1.4f, we obtain a value of 11.56 mm for t_E.

The lens surfaces, optical axis, principal planes, and points V_1, V_2, H_1, H_2, F_1, and F_2 are shown in Figure A1.4.

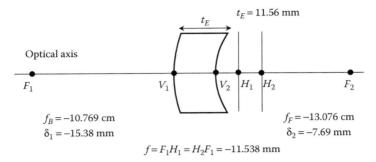

FIGURE A1.4 Negative meniscus lens surfaces, principal planes, and points V_{11}, V_{22}, H_{11}, H_{22}, F_{11}, and F_{22}.

1.5 a. Focal length f_2 is given by

$$f_2 = 10 f_1 \tag{A1.5a}$$

Substituting a value of 10 mm for f_1 in Equation A1.5a, we obtain a value of 100 mm for f_2.

   b. Distance d between the two lenses is given by

$$d = f_1 + f_2 \tag{A1.5b}$$

Substituting values of 10 mm for f_1 and 100 mm for f_2 in Equation A1.5b, we obtain a value of 110 mm for d.

   c. Diameter D_o of the output beam is given by

$$D_o = 10 D_i \tag{A1.5c}$$

where D_i is the diameter of the input beam. Substituting a value of 2 mm for D_i in Equation A1.5c, we obtain a value of 20 mm for D_o. A sketch of the beam expander is shown in Figure A1.5.

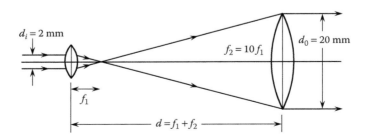

FIGURE A1.5 10x beam expander consisting of 10-mm and 100-mm focal length lenses.

1.6 a. Focal length f_2 is given by

$$f_2 = -10 f_1 \qquad\qquad \text{(A1.6a)}$$

Substituting a value of −10 mm for f_1 in Equation A1.6a, we obtain a value of 100 mm for f_2.

b. Distance d between the two lenses is given by

$$d = f_1 + f_2 \qquad\qquad \text{(A1.6b)}$$

Substituting values of −10 mm for f_1 and 100 mm for f_2 in Equation A1.6b, we obtain a value of 90 mm for d.

c. Diameter D_o of the output beam is given by

$$D_o = 10 D_i \qquad\qquad \text{(A1.6c)}$$

where D_i is the diameter of the input beam. Substituting a value of 2 mm for D_i in Equation A1.6c, we obtain a value of 20 mm for D_o. A sketch of the beam expander is shown in Figure A1.6.

1.7 a. Image distance s_i is given by

$$\frac{1}{s_i} = \frac{1}{f} - \frac{1}{s_o} \qquad\qquad \text{(A1.7a)}$$

where:

f is focal length of the lens
s_o is the distance of the object from the lens

Substituting values of 5.0 cm for f and 10.0 cm for s_o in Equation A1.7a, we obtain a value of 10.0 cm for s_i. The magnification is given by

$$M = -\frac{s_i}{s_o} \qquad\qquad \text{(A1.7b)}$$

Substituting values of 10.0 cm for s_i and 10.0 cm for s_o in Equation A1.7b, we obtain a value of −1 for the magnification M of the image.

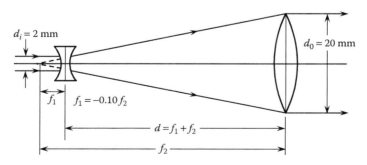

FIGURE A1.6 10x beam expander consisting of -10-mm and 100-mm focal length lenses.

1.8 a. The image distance s_i is given by

$$\frac{1}{s_i} = \frac{1}{f} - \frac{1}{s_o}$$ (A1.8a)

where:
   f is focal length of the lens
   s_o is the distance of the object from the lens

Substituting values of 5.0 cm for f and 20.0 cm for s_o in Equation A1.8a, we obtain a value of 6.7 cm for s_i. The magnification is given by

$$M = -\frac{s_i}{s_o}$$ (A1.8b)

Substituting values of 6.7 cm for s_i and 20.0 cm for s_o in Equation A1.8b, we obtain a value of −0.34 for the magnification M of the image.

1.9 a. 3-mm object is located 30 mm on the left of the lens of focal length f_1 equal to 25 mm. The image distance of the object s_{i1} formed by the f_1 lens is given by

$$\frac{1}{s_{i1}} = \frac{1}{f_1} - \frac{1}{s_{o1}}$$ (A1.9a)

Substituting values of 25.0 mm for f_1, and 30.0 mm for s_{o1} in Equation A1.9a, we obtain a value of 150 mm for s_{i1}. The magnification of this image is given by

$$M_1 = -\frac{s_{i1}}{s_{o1}}$$ (A1.9b)

Substituting values of 150 mm for s_{i1} and 30 mm for s_{o1} in Equation A1.9b, we obtain a value of −5.0 for M_1. The location of the virtual object relative to the f_2 lens is given by

$$s_{o2} = -(s_{i1} - d)$$ (A1.9c)

Substituting values of 150 mm for s_{i1} and 30 mm for the lens separation d in Equation A1.9c, we obtain a value of −120 mm for s_{o2}. The image distance due to the f_2 lens is given by

$$\frac{1}{s_{i2}} = \frac{1}{f_2} - \frac{1}{s_{o2}}$$ (A1.9d)

Substituting values of 40 mm for f_2 and −120 mm for s_{o2} in Equation A1.9d, we obtain a value of 30 mm for s_{i2}; this is location of the image of the object relative to the 40-mm focal length lens. The magnification provided by the f_2 lens is given by

$$M_2 = -\frac{s_{i2}}{s_{o2}} \qquad \text{(A1.9e)}$$

Substituting values of 30 mm for s_{i2} and −120 mm for s_{o2} in Equation A1.9e, we obtain a value of 0.25 for M_2. The total magnification of the two lenses is given by

$$M = M_1 M_2 \qquad \text{(A1.9f)}$$

Substituting values of −5 for M_1 and 0.25 for M_2 in Equation A1.9f, we obtain a value of −1.25 for M. The size of the image is given by

$$L_i = M L_0 \qquad \text{(A1.9g)}$$

Substituting values of −1.25 for M and 3 mm for L_0 in Equation A1.9g, we obtain a value of −3.75 mm for L_i.

1.10 a. The Sellmeier equation for fused silica (FS) is given by

$$n^2 = 1 + \frac{0.696166\lambda^2}{\lambda^2 - 0.004679} + \frac{0.407943\lambda^2}{\lambda^2 - 0.013512} + \frac{0.897479\lambda^2}{\lambda^2 - 97.93400} \qquad \text{(A1.10a)}$$

where λ is the wavelength in μm. A plot of n versus λ for FS is shown in Figure A1.7.

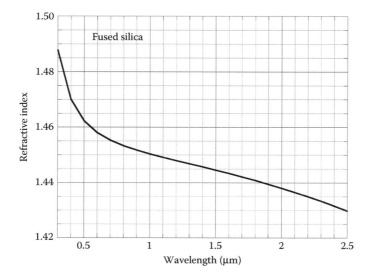

FIGURE A1.7   Refractive index of fused silica (FS).

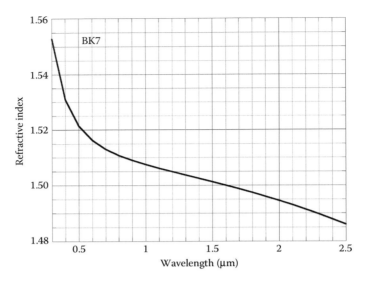

FIGURE A1.8   Refractive index of BK7.

b.   The Sellmeier equation for BK7 is given by

$$n^2 = 1 + \frac{1.039612\lambda^2}{\lambda^2 - 0.006001} + \frac{0.231792\lambda^2}{\lambda^2 - 0.020018} + \frac{1.010470\lambda^2}{\lambda^2 - 103.5607} \qquad \text{(A1.10b)}$$

A plot of n versus λ for BK7 is shown in Figure A1.8.

c.   The Sellmeier equation for calcium fluoride (CaF_2) is given by

$$n^2 = 1 + \frac{0.337601\lambda^2}{\lambda^2 - 0.000000} + \frac{0.701105\lambda^2}{\lambda^2 - 0.008775} + \frac{3.847815\lambda^2}{\lambda^2 - 1200.2} \qquad \text{(A1.10c)}$$

where λ is the wavelength in μm. A plot of n versus λ for for CaF_2 is shown in Figure A1.9.

d.   The Sellmeier equation for barium fluoride (BaF_2) is given by

$$n^2 = 1 + \frac{1.006307\lambda^2}{\lambda^2 - 0.000057} + \frac{0.143786\lambda^2}{\lambda^2 - 0.017520} + \frac{3.788478\lambda^2}{\lambda^2 - 2131.756} \qquad \text{(A1.10d)}$$

where λ is the wavelength in μm. A plot of n versus λ for for BaF_2 is shown in Figure A1.10.

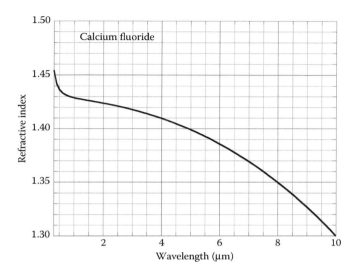

FIGURE A1.9   Refractive index of CaF$_2$.

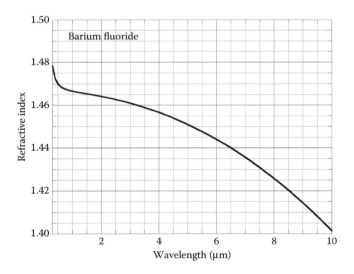

FIGURE A1.10   Refractive index of BaF$_2$.

e.   The Sellmeier equation for germanium (Ge) is given by

$$n^2 = 1 + \frac{14.75875\lambda^2}{\lambda^2 - 0.188629} + \frac{0.235256\lambda^2}{\lambda^2 - 1.593803} - \frac{24.88227\lambda^2}{\lambda^2 - 1695204} \qquad \text{(A1.10e)}$$

where λ is the wavelength in μm. A plot of n versus λ for Ge is shown in Figure A1.11.

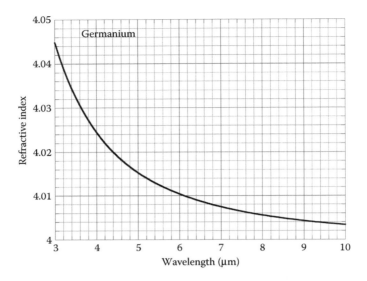

FIGURE A1.11   Refractive index of Ge.

1.11   Change in the focal length is given by

$$\Delta f = -\frac{\Delta n}{n-1} f \qquad \text{(A1.11)}$$

where Δn is the change in the refractive index of BK7 over the spectral region 0.4–0.7 μm. The values of refractive index of BK7 at 0.4 and 0.7 μm are 1.5308 and 1.5131, respectively. Therefore, the value of Δn is −0.0177. Substituting values of −0.0177 for Δn, 1.52 for n, and 25.000 mm for f in Equation A1.11, we obtain a value of 0.851 mm for Δf over the spectral range 0.4–0.7 μm. This represents a fractional change in the focal length $\Delta f/f$ of 3.4%.

1.12   Longitudinal spherical aberration for a plane parallel beam incident upon a plano-convex lens is given by

$$\text{LSA} \cong \frac{h^2}{8fn(n-1)} \left| \frac{n+2}{n-1} q^2 - 4(n+1)q + (3n+2)(n-1) + \frac{n^3}{n-1} \right| \qquad \text{(A1.12a)}$$

Shape factor is given by

$$q = \frac{R_2 + R_1}{R_2 - R_1} \qquad \text{(A1.12b)}$$

The value of q is 1.0 because R_2 is infinite. In this case, LSA is given by

$$\text{LSA} \cong \frac{h^2}{8fn(n-1)}\left[\frac{n+2}{n-1} - 4(n+1) + (3n+2)(n-1) + \frac{n^3}{n-1}\right] \qquad \text{(A1.12c)}$$

In our case, the values of h, f, and n are 23.6 mm, 85.0 mm, and 1.52, respectively. Using these values of h, f, and n in Equation A1.12, we obtain a value of 7.1 mm for LSA.

1.13 Radius of the comatic circle is given by

$$C_s(\text{cm}) = \frac{jh^3}{f^3}\left(\frac{3(2n+1)}{4n}p + \frac{3(n+1)}{4n(n-1)}q\right) \qquad \text{(A1.13)}$$

where $j = 5.0$ mm, $h = 12.5$ mm, $f = 25.0$ mm, $n = 1.5$, $p = -1$, and $q = 1$. Substituting these values of j, h, f, n, p, and q in Equation A1.13, we obtain a value of 0.313 mm or 313 μmm for C_s.

1.14 Shape factor for a coma-free lens for a distant off-axis point object is given by

$$q = \frac{(2n+1)(n-1)}{n+1} \qquad \text{(A1.14)}$$

The value of refractive index n for germanium at 6 μm is 4.01. Using this value of n in Equation A1.14, we obtain a value of 4.82 for q.

1.15 The separation between the image distances in the sagittal and tangential planes is given by

$$s = f\sin^2\phi \qquad \text{(A1.15)}$$

where:
  f is the focal length of the lens
  ϕ is the angle of incidence of the chief ray

Substituting values of 50 mm for f and 10° for ϕ in Equation A1.15, we obtain a value of 1.5 mm for s.

1.16 Effective focal length of the combination of two thin lenses of focal lengths f_1 and f_2 separated by a distance d is given by

$$\frac{1}{f} = \frac{1}{f_1} + \frac{1}{f_2} - \frac{d}{f_1f_2} \qquad \text{(A1.16a)}$$

If the Petzval surface is planar, f_1 and f_2 related by

$$f_1 + f_2 = 0 \qquad \text{(A1.16b)}$$

assuming that the two lenses have the same refractive index. Equation A1.16b yields the condition that $f_2 = -f_1$. Using this condition, Equation A1.16a yields the result

$$f = \frac{f_1^2}{d} \tag{A1.16c}$$

Using the value 30 mm for f_1 and 30 mm for the separation d in Equation A1.16c, we obtain a value of 30 mm for f.

1.17 Magnification of a telescope is given by

$$M_T = \frac{f_O}{f_E} \tag{A1.17}$$

Substituting values of 200 cm for f_O and 5.0 cm for f_E in Equation A1.17, we obtain a value of 40 for M_T.

1.18 Depth of focus is given by

$$\text{DOF} \approx 2(f\#)\delta\left(\frac{s_o}{f}\right)^2 \tag{A1.18}$$

Substituting values of 1.8 for $f\#$, 25 μm for δ, 250 mm for s_o, and 50 mm for f in Equation A1.18, we obtain a value of 2.25 mm for *DOF*.

1.19 a. Focal length of the combination of L_1 and L_2 lenses in contact is given by

$$\frac{1}{F_{L1L2}} = \frac{1}{f_{L1}} + \frac{1}{f_{L2}} \tag{A1.19a}$$

Substituting values of 200 mm for f_{L1} and −62.5 mm for f_{L2} in Equation A1.19a, we obtain a value of −90.9 mm for F_{L1L2}. Focal length of the combination of L_3 and L_0 in contact is given by

$$\frac{1}{F_{L3L0}} = \frac{1}{f_{L3}} + \frac{1}{f_{L0}} \tag{A1.19b}$$

Substituting values of 200 mm for f_{L3} and 100 mm for f_{L0} in Equation A1.19b, we obtain a value of 66.7 mm for f_{L3L0}. Focal length of the zoom lens is given by

$$\frac{1}{F_Z} = \frac{1}{f_{L1L2}} + \frac{1}{f_{L3L0}} - \frac{d}{f_{L1L2}f_{L3L0}} \tag{A1.19c}$$

Substituting values of −90.9 mm for f_{L1L2}, 66.7 mm for f_{L3L0}, and 120 mm for d in Equation A1.19c, we obtain a value of 42.0 mm for F_Z. The value of the back focal length is equal to 97.4 mm.

b. Focal length of the combination of L_2 and L_3 and L_0 lenses in contact is given by

$$\frac{1}{F_{L2L3L0}} = \frac{1}{f_{L2}} + \frac{1}{f_{L3}} + \frac{1}{L_0} \tag{A1.19d}$$

Substituting values of −62.5 mm for f_{L2}, 200 mm for f_{L3}, and 100 mm for L_0 in Equation A1.19d, we obtain a value of −1,000 mm for F_{L2L3L0}. Focal length of the zoom lens is given by

$$\frac{1}{F_Z} = \frac{1}{f_{L1}} + \frac{1}{f_{L2L3L0}} - \frac{d}{f_{L1}f_{L2L3L0}} \tag{A1.19e}$$

Substituting values of 200 mm for f_{L1}, −1,000 mm for f_{L2L3L0}, and 120 mm for the separation d in Equation A1.19e, we obtain a value of 217.5 mm for F_Z. The value of the back focal length is 87.0 mm.

1.20 Values of the effective focal length f_z and the back focal length f_{Bz} of the zoom lens are calculated using the following equations:

$$\frac{1}{f_{1,2}} = \frac{1}{200} - \frac{1}{62.5} + \frac{d_{1,2}}{200*62.5} \tag{A1.20a}$$

$$f_{B1,2} = f_{1,2}\left(1 - \frac{d_{1,2}}{200}\right) \tag{A1.20b}$$

$$d_{1,2-3} = 120 + (f_{1,2} - f_{B1,2} - d_{1,2}) \tag{A1.20c}$$

$$\frac{1}{f_z} = \frac{1}{f_{1,2}} + \frac{1}{66.67} - \frac{d_{1,2-3}}{f_{1,2}*66.67} \tag{A1.20d}$$

$$f_{Bz} = f_z\left(1 - \frac{d_{1,2-3}}{f_{1,2}}\right) \tag{A1.20e}$$

Calculated values of the effective focal length f_z and back focal length f_{Bz} of the zoom lens are listed in Table A1.1.

A plot of f_z versus $d_{1,2}$ is shown in Figure A1.12.

TABLE A1.1
Effective Focal Length f_z and
Back Focal f_{Bz}

$d_{1,2}$	F_z	f_{Bz}
0	42.0	97.5
10	47.9	99.2
20	55.0	101.1
30	63.5	103.1
40	73.8	105.0
50	86.2	106.9
60	101.2	108.5
70	119.1	109.5
80	139.7	109.5
90	162.4	107.8
100	185.2	103.7
110	204.9	96.7
120	217.4	87.0

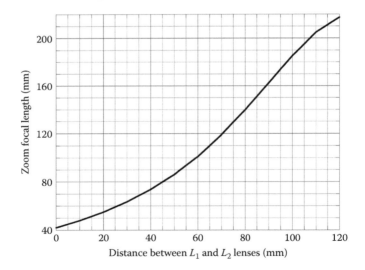

FIGURE A1.12 Plot of the zoom focal length.

1.21 Image of the object formed by the f_1 lens is located at a distance of s_{i1}, which is given by

$$\frac{1}{s_{i1}} = \frac{1}{f_1} - \frac{1}{s_{o1}} \qquad (A1.21a)$$

Substituting values of 30 mm for f_1 and 70 mm for s_{o1} in Equation A1.21a, we obtain a value of 52.5 mm for s_{i1}.

The length of the image L_{i1} is given by

$$L_{i1} = -L_o \frac{s_{i1}}{s_{o1}}$$ (A1.21b)

Substituting values of 5 mm for L_o, 52.5 mm for s_{i1} and 70 mm for s_{o1} in Equation A1.21b, we obtain a value of -3.75 mm for L_{i1}.

The image formed by the f_1 lens is located at a distance of s_{o2} equal to -7.5 mm from the f_2 lens. The image formed by the f_2 lens is located at a distance s_{i2} given by

$$\frac{1}{s_{i2}} = \frac{1}{f_2} - \frac{1}{s_{o2}}$$ (A1.21c)

Substituting values of 15 mm for f_2 and -7.5 mm for s_{o2} in Equation A1.21c, we obtain a value of 5.0 mm for s_{i2}. The length of the image L_{i2} formed by the f_2 lens is given by

$$L_{i2} = -L_{i1} \frac{s_{i2}}{s_{o2}}$$ (A1.21d)

Substituting values of -3.75 mm for L_{i1}, 5 mm for s_{i2}, and -7.5 mm for s_{o2} in Equation A1.21d, we obtain a value of -2.5 mm for L_{i2}.

Thus, the length of the image of the object formed by the relay telescope is -2.5 mm; it implies that the image is inverted. This image is located at a distance of 5 mm on the right-hand side of the relay telescope.

2.1 The focal length for rays, which are incident at a height h from the axis of a concave spherical mirror with radius of curvature R is given by

$$f_h = x_1 + x_2$$ (A2.1a)

$$x_1 = R - \sqrt{R^2 - h^2}$$ (A2.1b)

$$x_2 = h\cot(2\theta_i)$$ (A2.1c)

with

$$\theta_i = \sin^{-1}\left(\frac{h}{R}\right)$$ (A2.1d)

Substituting values of 50 mm for R and 15 mm for h in Equation A2.1b, we obtain a value of 2.30 mm for x_1. Equations A2.1c and A2.1d yield a value of 21.49 mm for x_2. Substituting these values of x_1 and x_2 in Equation A2.1a, we obtain a value of 23.79 mm for f_h, which is less than the value of 25.0 mm for the paraxial focal length f_p, as expected.

The change in the focal length is given by

$$\Delta f = f_h - f_p \tag{A2.1e}$$

Substituting values of 23.79 mm for f_h and 25.00 mm for f_p in Equation A2.1e, we obtain a value of -1.21 mm for Δf. The fractional change in the focal length is given by $\Delta f / f_p = -4.84\%$.

2.2   The slope of the tangent to the parabola is given by

$$m = \frac{h}{R} = \tan \theta_i \tag{A2.2a}$$

where θ_i is the angle of incidence. Using the values of 15 mm for h and 50 mm for R in Equation A2.2a, we obtain a value of 0.30 for $\tan \theta_i$, which yields a value of 16.7° for θ_i. The focal length for a ray incident at a height h from the optical axis is given by

$$f_h = x_1 + x_2 \tag{A2.2b}$$

$$x_1 = \frac{h^2}{2R} \tag{A2.2c}$$

$$x_2 = \frac{h}{\tan 2\theta_i} \tag{A2.2d}$$

Using values of 15 mm for h and 50 mm for R in Equation A2.2c, we obtain values of 2.25 mm for x_1. Using values of 15 mm for h and 16.7° for θ_i in Equation A2.2d, we obtain a value of 22.75 mm for x_2. Using these values of x_1 and x_2 in Equation A2.2b, we obtain a value of 25.00 mm for f_h, which is exactly equal to the value of 25.00 mm for the paraxial focal length f_p.

2.3   The image distance is given by

$$\frac{1}{s_i} = \frac{1}{f} - \frac{1}{s_o} \tag{A2.3a}$$

which yields the result

$$s_i = \frac{s_o f}{s_o - f} \tag{A2.3b}$$

Using values of 40 mm for s_o and 25 mm for f in Equation A2.3b, we obtain a value of 66.7 mm for s_i. The length of the image is given by

$$l_i = -\frac{s_i}{s_o} l_o \tag{A2.3c}$$

Substituting values of 66.7 mm for s_i, 40 mm for s_o, and 3.0 mm for l_o in Equation A2.3c, we obtain a value of −5.0 mm for for l_i.

2.4   The diameter of the secondary mirror that is illuminated is given by

$$D_S = D_P \left(1 - \frac{d}{f_P} \right) \tag{A2.4}$$

where:

D_P is the diameter of the primary mirror

d is the distance between the primary and secondary mirrors

f_P is the focal length of the primary mirror of the Hubble Space Telescope (HST)

Using values of 2.40 m for D_P, 4.91 m for the separation d, and 5.52 m for f_P in Equation A2.4, we obtain a value of 0.265 m for D_S.

2.5   The angle of incidence is given by

$$\theta_i = \tan^{-1} \left(\frac{h}{R} \right) - \theta_o \tag{A2.5a}$$

where θ_o is the angle of the incident beam with the axis of the parabolic mirror. Substituting values of 15 mm for h, 50 mm for R, and 5.0° for θ_o in Equation A2.5a, we obtain a value of 11.7° for θ_i. The focal length for a ray incident at a height h from the axis of the mirror is given by

$$f_h = \frac{x_1 + x_2}{\cos \theta_o} \tag{A2.5b}$$

$$x_1 = \frac{h^2}{2R} \tag{A2.5c}$$

$$x_2 = \frac{h}{\tan \theta_o + \tan(2\theta_i + \theta_o)} \tag{A2.5d}$$

Substituting values of 15 mm for h and 50 mm for R in Equation A2.5c, we obtain a value of 2.25 mm for x_1. Substituting values of 15 mm for h, 5° for θ_o, and 11.7° for θ_i in Equation A2.5d, we obtain a value of 23.88 mm for x_2. Substituting values of 2.25 mm for x_1, 23.88 mm for x_2, and 5° for θ_o in Equation A2.5b, we obtain a value of 26.23 mm for f_h.

2.6   Effective focal length of the HST is given by

$$\frac{1}{f} = \frac{1}{f_P} + \frac{1}{f_S} - \frac{d}{f_P f_S} \tag{A2.6a}$$

which yields the relationship

$$f = \frac{f_P f_S}{f_P + f_S - d} \tag{A2.6b}$$

Using values of 5.52 m for f_P, −0.679 m for f_S, and 4.91 m for the separation d in Equation A2.6b, we obtain a value 54.32 m for f.

The back focal length of the HST is given by

$$f_B = f\left(1 - \frac{d}{f_P}\right) \tag{A2.6c}$$

Using values of 54.32 m for f, 4.91 m for the separation d, and 5.52 m for f_P in Equation A2.6c, we obtain a value of 6.00 m for f_B, which is the distance of the focal spot of the HST from the secondary mirror.

2.7 The value of the separation between the primary and secondary mirrors of the HST is given by

$$d = f_P + f_S - \frac{f_P f_S}{f} \tag{A2.7}$$

Substituting values of 5.52 m for f_P, −0.679 m for f_S, and 57.6 m for f in Equation A2.7, we obtain a value of 4.91 m for the separation d.

2.8 a and b of the Subaru Telescope primary mirror are given by

$$a_P = -\frac{2 f_P}{(K_P + 1)} \tag{A2.8a}$$

$$b_P = \sqrt{2 a_P f_P} \tag{A2.8b}$$

Using values of 15 m for f_P and −1.00835051 for K_P in Equation A2.8a, we obtain a value of 3592.6 m for a_P. Using this value of 3,592.6 m for a_P and 15.0 m for f_P in Equation A2.8b, we obtain a value of 328.3 m for b_P.

a and b of the Subaru Telescope secondary mirrors are given by

$$a_S = -\frac{2 f_S}{(K_S + 1)} \tag{A2.8c}$$

$$b_S = \sqrt{2 a_S f_S} \tag{A2.8d}$$

Cassegrain Optical: Using values of −2.762 m for f_S and −1.917322232 for K in Equation A2.8c, we obtain a value of −6.02 m for a_S. Using this value of −6.02 m for a_S and −2.762 m for f_S in Equation A2.8d, we obtain a value of 5.77 m for b_S.

Nasmyth Optical: Substituting values of −2.939 m for f_S and −1.865055214 for K in Equation A2.8c, we obtain a value of −6.80 m for a_S. Substituting this value −6.80 m for a_S and −2.939 m for f_S in Equation A2.8d, we obtain a value of 6.32 m for b_S.

Cassegrain IR: The values of f_S and K_S are the same as for the Cassegrain Optical. Therefore, the values of a_S and b_S are also the same as for the Cassegrain optical.

2.9   The focal length of the Subaru Telescope is given by

$$f = \frac{f_P f_S}{f_P + f_S - d}$$ (A2.9)

Cassegrain Optical: Substituting values of 15.0 m for f_P and −2.762 m for f_S, and 12.652 m for the separation d in Equation A2.9, we obtain a value of 100.1 m for f.

Nasmyth Optical: Substituting values of 15 m for f_P and −2.939 m for f_S, and 12.484 m for the separation d in Equation A2.9, we obtain a value of 104.2 m for f.

Cassegrain IR: f is equal to 100.1 m, the same as for the Cassegrain Optical because the values of f_P, f_S, and d are the same in the two cases.

2.10   The back focal length is given by

$$f_B = f\left(1 - \frac{d}{f_P}\right)$$ (A2.10)

Cassegrain Optical: Using the values of 100.1 m for f, 12.652 m for the separation d, and 15 m for f_P in Equation A2.10, we obtain a value of 15.67 m for f_B.

Nasmyth Optical: Using the values of 104.2 m for f, 12.484 m for the separation d, and 15 m for f_P in Equation A2.10, we obtain a value of 17.48 m for f_B.

Cassegrain IR: The value of f_B is equal to 15.67 m, same as for the Cassegrain Optical because the values of f, d, and f_P are the same in the two cases.

2.11   The focal lengths f_1 and f_2 of an ellipsoidal mirror are given by

$$f_1 = a - \sqrt{a^2 - b^2}$$ (A2.11a)

$$f_2 = a + \sqrt{a^2 - b^2}$$ (A2.11b)

The magnification for the source located at f_2 is given by

$$M = \frac{f_1}{f_2}$$ (A2.11c)

Substituting values of 1/6 for M and 25 mm for f_1 in Equation A2.11c, we obtain a value of 150 mm for f_2. Using values of 25 mm for f_1 and 150 mm for f_2 in Equations A2.11a and A2.11b, we obtain values of 87.5 mm for a and 61.2 mm for b. The conic constant is given by

$$K = -1 + \frac{b^2}{a^2}$$ (A2.11d)

Using the values of 87.5 mm for a and 61.2 mm for b in Equation A2.11d, we obtain a value of -0.515 equal to $-25/49$ for K.

3.1   The grating groove spacing d is given by

$$d = \frac{m\lambda}{(\sin\theta_m - \sin\theta_i)} \tag{A3.1a}$$

Substituting values of 2 for the order m, 0.5145 μm for λ, 55.4° for θ_m, and 0° for θ_i in Equation A3.1a, we obtain a value of 1.25 μm for the groove spacing d. The number of grating grooves is given by

$$N = \frac{1}{d} \tag{A3.1b}$$

Substituting a value of 1.25 μm for the groove spacing d in Equation A3.1b, we obtain a value of 800 grooves/mm for N.

3.2   The diffraction order m is given by

$$m = \frac{d(\sin\theta_m - \sin\theta_i)}{\lambda} \tag{A3.2a}$$

The maximum value of $\sin\theta_m$ is less than or equal to 1. Therefore, the maximum value of m is given by

$$m \leq \frac{d(1 - \sin\theta_i)}{\lambda} \tag{A3.2b}$$

$$d = \frac{1}{N} \tag{A3.2c}$$

Substituting a value of 500 grooves/mm for N in Equation A3.2c, we obtain a value of 2.0 μm for the groove spacing d. Using this value of 2.0 μm for the groove spacing d, 20° for θ_i, and 0.532 μm for λ in Equation A3.2b, m is given by

$$m \leq 2.47 \tag{A3.2d}$$

Hence the maximum value of m is 2 because m is an integer.

3.3   The angle of diffraction is given by

$$\theta_m = \sin^{-1}\left(\sin\theta_i + \frac{m\lambda}{d}\right) \tag{A3.3a}$$

$$d = \frac{1}{N} \tag{A3.3b}$$

Using the value of 1200 grooves/mm for N in Equation A3.3b, we obtain a value of 0.833 μm for the groove spacing d. Using this value of

0.833 μm for the groove spacing d, 15° for θ_i, and 1 for order m, Equation A3.3a for the angle of diffraction becomes

$$\theta_m = \sin^{-1}(0.259 + 1.2\lambda) \tag{A3.3c}$$

Substituting the value of 0.4358 μm for λ of the blue line in Equation A3.3c, we obtain a value of 51.4° for θ_m for the blue line. Substituting the value of 0.5461 μm for λ of the green line in Equation A3.3c, we obtain a value of 66.1° for θ_m for the green line. Therefore, the angular separation between the blue and green diffracted beams is 14.7°.

3.4 The angle of diffraction is given by

$$\theta_m = \sin^{-1}\left(\sin\theta_i + \frac{m\lambda}{d}\right) \tag{A3.4a}$$

$$d = \frac{1}{N} \tag{A3.4b}$$

Using the value of 500 grooves/mm for N in Equation A3.4b, we obtain a value of 2.00 μm for the groove spacing d. For this value of 2.00 μm for the groove spacing d, 0° for θ_i, and 1 for order m, Equation A3.4a for the angle of diffraction becomes

$$\theta_m = \sin^{-1}(0.5\lambda) \tag{A3.4c}$$

where λ is in μm. Using a value of 0.550 μm for λ in Equation A3.4c, we obtain a value of 16.0° for θ_m. Using a value of 0.560 μm for λ, we obtain a value of 16.3° for θ_m. Therefore, the angle $\Delta\theta_m$ between the diffracted 0.550 and 0.560 μm beams is equal to 0.3°. The separation between the 0.550 and 0.560 μm focal spots in the focal plane of a lens of focal length f is given by

$$S = f\Delta\theta_m \tag{A3.4d}$$

Using values of 1.0 m for f and 0.3° for $\Delta\theta_m$ in Equation A3.4d, we obtain a value of 5.2 mm for S.

3.5 The resolving power of the grating is given by

$$R = Nm \tag{A3.5a}$$

where N is the total number of grating grooves illuminated by the incident light. N is given by

$$N = \frac{w}{d} \tag{A3.5b}$$

where w is the grating width that is illuminated by the incident light beam. Substituting values of 50.0 mm for w and 1200 grooves/mm for

$1/d$ in Equation A3.5b, we obtain a value of 6×10^4 for N. Substituting this value of 6×10^4 for N and 2 for order m, we obtain a value of 1.2×10^5 for R.

The spectral resolution is given by

$$\Delta\lambda = \frac{\lambda}{R} \tag{A3.5c}$$

Substituting values of 500 nm for λ and 1.2×10^5 for R in Equation A3.5c, we obtain a value of 0.0042 nm for $\Delta\lambda$.

3.6   The resolving power of the grating is given by

$$R = \frac{\lambda}{\Delta\lambda} \tag{A3.6a}$$

Using values of 500 nm for λ and 0.01 nm for $\Delta\lambda$ in Equation A3.6a, we obtain a value of 5×10^4 for R. The number of grating grooves is given by

$$N = \frac{R}{m} \tag{A3.6b}$$

Using values of 5×10^4 for R and 3 for order m in Equation A3.6a in A3.6b, we obtain a value of 1.667×10^4 for N.

3.7   TE-polarization corresponds to E-field parallel to the grating grooves. Therefore, the efficiency for the TE-polarization at 600 nm is 93%.

TM-polarization corresponds to E-field perpendicular to the grating grooves. Therefore, the efficiency for the TM-polarization at 600 nm is 50%.

3.8   Angle of diffraction θ_m is given by

$$\sin\theta_m = \sin\theta_i + m\frac{\lambda}{d} \tag{A3.8a}$$

If diffracted light of wavelength λ_1 in the m order overlaps diffracted light of wavelength $\lambda_2 < \lambda_1$ in the $m + 1$ order, then we have

$$\sin\theta_m = \sin\theta_i + m\frac{\lambda_1}{d} = \sin\theta_i + (m+1)\frac{\lambda_2}{d} \tag{A3.8b}$$

This yields the condition

$$m\lambda_1 = (m+1)\lambda_2 \tag{A3.8c}$$

Substituting values of 630 nm for λ_1 and 420 nm for λ_2 in Equation A3.8c, we obtain a value of 2 for order m. Therefore, the values of the two diffraction orders are 2 and 3 for wavelengths of 630 and 420 nm, respectively.

3.9   The resolving power is given by

$$R = Nm \tag{A3.9a}$$

Substituting values of 1×10^6 for R and 3 for order m in Equation A3.9a, we obtain a value of 3.33×10^5 grooves for N. The width of the grating is given by

$$w = \frac{N}{G_D} \tag{A3.9b}$$

where G_D is the groove density of the grating. Substituting values of 3.33×10^5 grooves for N and 1800 grooves/mm for G_D in Equation A3.9b, we obtain a value of 185 mm for w.

4.1   The fraction of the unpolarized light that is transmitted through the combination of two polarizers is given by

$$\frac{I_T}{I_0}(\theta) = T_1 T_2 (\cos \theta)^2 \tag{A4.1a}$$

where:

T_1 is the transmittance of the first polarizer for unpolarized light
T_2 is the transmittance of the second polarizer for polarized light

Using values of 45% for T_1 and 90% for T_2 in Equation A4.1a, the fraction of the unpolarized light that is transmitted through the combination of two polarizers is given by

$$\frac{I_T}{I_0}(\theta) = 0.405 (\cos \theta)^2 \tag{A4.1b}$$

Substituting values of 30°, 45°, and 60° for θ in Equation A4.1b, we obtain values of 30.4%, 20.3%, and 10.1%, respectively, for the fraction of unpolarized light that is transmitted through the combination of two polarizers.

4.2   The intensity of the light transmitted through the linear polarizer is given by

$$I_T(\theta) = I_0 (\cos \theta)^2 \tag{A4.2a}$$

where I_0 is the intensity of light transmitted through the linear polarizer for θ equal to 0°. We note that

$$\cos(\theta) = \cos(-\theta) \tag{A4.2b}$$

Therefore, the intensity of the light transmitted through the polarizer is modulated at twice the frequency at which the polarizer is rotated about an axis, which is perpendicular to the plane of the polarizer. Using a value of 50 cycles/s for the frequency of rotation of the polarizer, the intensity of the light transmitted through the polarizer is modulated at 100 cycles/s.

4.3   The angle of incidence θ_2 at the second surface of the plane-parallel plate is given by

$$\sin\theta_2 = \frac{1}{n}\sin\theta_1 \qquad\qquad\text{(A4.3a)}$$

where θ_1 is the angle of incidence at the first surface of the plane-parallel plate. For $\theta_1 = \theta_B$ and $\tan\theta_B = n$, we obtain

$$\theta_2 = 90° - \theta_B \qquad\qquad\text{(A4.3b)}$$

The Brewster angle θ_{B2} at the second surface of the plane-parallel plate is given by

$$\tan\theta_{B2} = \frac{1}{n} = \cot\theta_B = \tan(90° - \theta_B) = \tan\theta_2 \qquad\text{(A4.3c)}$$

which shows that

$$\theta_2 = \theta_{B2} \qquad\qquad\text{(A4.3d)}$$

as required.

4.4   The degree of polarization of light transmitted by a pile-of-plates polarizer is given by

$$P = \frac{m}{m + \left(\dfrac{2n}{n^2 - 1}\right)^2} \qquad\qquad\text{(A4.4)}$$

where:
  m is the number of plates
  n is the refractive index of the plate material

Using values of 5 for m and 4.0 for n in Equation A4.4, we obtain a value of 94.6% for P.

4.5   The critical angle θ_c for total internal reflection is given by

$$\sin\theta_c = \frac{1}{n} \qquad\qquad\text{(A4.5a)}$$

Using a value of 45° for θ_c in Equation A4.5a, we obtain a value of 1.414 for n.

The Brewster angle is given by

$$\theta_B = \tan^{-1} n \qquad\qquad\text{(A4.5b)}$$

Using a value of 1.414 for n in Equation A4.5b, we obtain a value of 54.7° for θ_B.

4.6   The phase changes in the p and s polarization components upon total internal reflection are given by

$$\tan\frac{\delta_p}{2} = \frac{n\sqrt{n^2\sin^2\theta_i}}{\cos\theta_i} \qquad\text{(A4.6a)}$$

$$\tan\frac{\delta_s}{2} = \frac{\sqrt{n^2\sin^2\theta_i}}{n\cos\theta_i} \qquad\text{(A4.6b)}$$

The phase difference between the p and s components is given by

$$\delta = \delta_p - \delta_s \qquad\text{(A4.6c)}$$

A plot of the phase difference δ versus θ_i for the CsI Fresnel rhomb with n equal to 1.7 at 32 µm is shown in Figure A4.1.

The above plot shows that the phase difference δ is equal to 45° for angle of incidence θ_i equal to 38.6°.

4.7   The thickness of the quarter-wave plate is given by

$$d = \frac{\lambda}{4\Delta n} \qquad\text{(A4.7a)}$$

$$\Delta n = n_E - n_O \qquad\text{(A4.7b)}$$

Substituting values of 1.6584 for n_E and 1.4864 for n_O in Equation A4.7b, we obtain a value of 0.172 for Δn. Substituting values of 1.06 µm for λ and 0.172 for Δn in Equation A4.7a, we obtain a value of 1.541 µm for the thickness d.

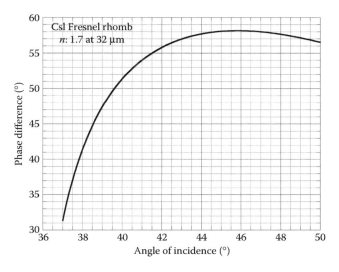

FIGURE A4.1   Phase difference versus angle of incidence.

4.8   The Jones matrix for a quarter-wave plate at an angle θ with the x-axis is given by

$$M_{\text{QWP}} = \begin{bmatrix} \cos^2\theta + i\sin^2\theta & (1-i)\sin\theta\cos\theta \\ (1-i)\sin\theta\cos\theta & \sin^2\theta + i\cos^2\theta \end{bmatrix} \tag{A4.8a}$$

The Jones matrix for a half-wave plate at an angle θ with the x-axis is given by

$$M_{\text{HWP}} = \begin{bmatrix} \cos^2\theta + i\sin^2\theta & (1-i)\sin\theta\cos\theta \\ (1-i)\sin\theta\cos\theta & \sin^2\theta + i\cos^2\theta \end{bmatrix}\begin{bmatrix} \cos^2\theta + i\sin^2\theta & (1-i)\sin\theta\cos\theta \\ (1-i)\sin\theta\cos\theta & \sin^2\theta + i\cos^2\theta \end{bmatrix}$$

$$\tag{A4.8b}$$

Using matrix multiplication, the above equation reduces to

$$M_{\text{HWP}} = \begin{bmatrix} \cos^2\theta - \sin^2\theta & 2\sin\theta\cos\theta \\ 2\sin\theta\cos\theta & \sin^2\theta - \cos^2\theta \end{bmatrix} \tag{A4.8c}$$

Substituting a value of $\theta = 0°$ in Equation A4.8c, we obtain

$$M_{\text{HWP}} = \begin{bmatrix} 1 & 0 \\ 0 & -1 \end{bmatrix} \tag{A4.8d}$$

Substituting a value of $\theta = 45°$ in Equation A4.8c, we obtain

$$M_{\text{HWP}} = \begin{bmatrix} 0 & 1 \\ 1 & 0 \end{bmatrix} \tag{A4.8e}$$

4.9   The Jones matrix for the right circular polarizer is given by

$$M_{\text{RCP}} = M_{\text{QWP}}(0°)\,M_{\text{LP}}(45°) \tag{A4.9a}$$

$$M_{\text{QWP}}(0°) = \begin{bmatrix} 1 & 0 \\ 0 & i \end{bmatrix} \tag{A4.9b}$$

$$M_{\text{LP}}(45°) = \frac{1}{2}\begin{bmatrix} 1 & 1 \\ 1 & 1 \end{bmatrix} \tag{A4.9c}$$

Using the above equations for $M_{QWP}(90°)$ and $M_{LP}(45°)$, the Jones matrix for the right circular polarizer is given by

$$M_{RCP} = \frac{1}{2}\begin{bmatrix} 1 & 1 \\ i & i \end{bmatrix} \tag{A4.9d}$$

4.10 The Jones matrix for the combination of the quarter-wave plate and the linear polarizer is given by

$$M = M_{LP}(45°)\, M_{QWP}(0°) = \frac{1}{2}\begin{bmatrix} 1 & i \\ 1 & i \end{bmatrix} \tag{A4.10a}$$

The Jones vector for the incident light may be taken as

$$J_{inc} = \frac{1}{\sqrt{2}}\begin{bmatrix} 1 \\ 1 \end{bmatrix} \tag{A4.10b}$$

The Jones vector for the transmitted light is then given by

$$J_{trans} = MJ_{inc} = \frac{1}{2\sqrt{2}}\begin{bmatrix} 1 & i \\ 1 & i \end{bmatrix}\begin{bmatrix} 1 \\ 1 \end{bmatrix} = \frac{1+i}{2\sqrt{2}}\begin{bmatrix} 1 \\ 1 \end{bmatrix} \tag{A4.10c}$$

which is the Jones vector for linearly polarized light. This shows that we obtain circularly polarized light only by sending the unpolarized light in the right order through a linear polarizer and a quarter-wave plate.

4.11 The Jones matrix for the linear polarizer with its transmission axis at 45° is given by

$$M_{LP}(45°) = \frac{1}{2}\begin{bmatrix} 1 & 1 \\ 1 & 1 \end{bmatrix} \tag{A4.11a}$$

The Jones matrix for the linear polarizer with its transmission axis at 90° is given by

$$M_{LP}(90°) = \begin{bmatrix} 0 & 0 \\ 0 & 1 \end{bmatrix} \tag{A4.11b}$$

The Jones matrix for the combination of these two linear polarizers is given by

$$M = \frac{1}{2}\begin{bmatrix} 0 & 0 \\ 0 & 1 \end{bmatrix}\begin{bmatrix} 1 & 1 \\ 1 & 1 \end{bmatrix} = \frac{1}{2}\begin{bmatrix} 0 & 0 \\ 1 & 1 \end{bmatrix} \tag{A4.11c}$$

The Jones vector for the incident light is

$$J_i = \begin{bmatrix} 1 \\ 0 \end{bmatrix}$$ (A4.11d)

The Jones vector for the transmitted light is given by

$$J_t = MJ_i = \frac{1}{2}\begin{bmatrix} 0 & 0 \\ 1 & 1 \end{bmatrix}\begin{bmatrix} 1 \\ 0 \end{bmatrix} = \frac{1}{2}\begin{bmatrix} 0 \\ 1 \end{bmatrix}$$ (A4.11e)

which represents linearly polarized light in the vertical direction with E-field amplitude equal to ½ of that of incident light. This implies that intensity of the transmitted light is ¼ of that of the incident light as expected.

5.1 The minimum germanium window thickness is given by

$$t_w = \left(\frac{1.25\Delta P_w S_F}{M_R}\right)^{1/2} R_w$$ (A5.1)

where:

ΔP_w = pressure differential on the window = 150 psi = 1.031×10^6 Pa

S_F = safety factor = 6

M_R = modulus of rupture = 90 MPa = 9.0×10^7 Pa

R_w = radius of the window = 25 mm

Using these values of P_w, S_F, M_R, and R_w, we obtain a value of 7.3 mm for the minimum thickness of the Ge window.

5.2 This problem for the CaF$_2$ window is similar to the previous problem 5.1 for the Ge window. The variables for the CaF$_2$ window are ΔP_w = pressure differential on the window = 15 psi = 1.031×10^5 Pa, S_F = safety factor = 4, M_R = modulus of rupture = 37 MPa = 3.7×10^7 Pa, and R_w = radius of the window = 12.5 mm. Using these values of the variables for the CaF$_2$ window, we obtain a value of 1.5 mm for the minimum thickness of the CaF$_2$ window.

5.3 The single-surface reflectance is given by

$$R = \left(\frac{n-1}{n+1}\right)^2$$ (A5.3a)

Using a value of 4.00 for n of Ge window at 10.6 μm, we obtain a value of 0.36 for R of Ge window at 10.6 μm.

The transmittance of the uncoated absorption-free window is given by

$$T_E = \frac{1-R}{1+R}$$ (A5.3b)

The reflection loss of the uncoated window is given by

$$L_R = 1 - T_E = \frac{2R}{1+R} \qquad \text{(A5.3c)}$$

Using a value of 0.36 for R, we obtain a value of 0.53 = 53% for the reflection loss L_R of the uncoated Ge window at 10.6 µm.

5.4   The single-surface reflectance is given by

$$R = \left(\frac{n-1}{n+1}\right)^2 \qquad \text{(A5.4a)}$$

Using a value of 2.37 for n of KRS-5 window at 10.6 µm, we obtain a value of 0.165 for R of KRS-5 window at 10.6 µm.

The external transmittance of the uncoated absorption-free window is given by

$$T_E = \frac{1-R}{1+R} \qquad \text{(A5.4b)}$$

The reflection loss of the uncoated window is given by

$$L_R = 1 - T_E = \frac{2R}{1+R} \qquad \text{(A5.4c)}$$

Using a value of 0.165 for R, we obtain a value of 0.283 = 28.3% for the reflection loss L_R of the uncoated KRS-5 window at 10.6 µm.

5.5   The single-surface reflectance is given by

$$R = \left(\frac{n-1}{n+1}\right)^2 \qquad \text{(A5.5a)}$$

Using a value of 1.457 for n of UV grade fused silica window at 632 nm in Equation A5.5a, we obtain a value of 0.035 for R of the UV grade fused silica window at 632 nm.

The external transmittance of the uncoated absorption-free window is given by

$$T_E = \frac{1-R}{1+R} \qquad \text{(A5.5b)}$$

The reflection loss of the uncoated window is given by

$$L_R = 1 - T_E = \frac{2R}{1+R} \qquad \text{(A5.5c)}$$

Using a value of 0.035 for R in Equation A5.5c, we obtain a value of 6.8% for the reflection loss L_R of the uncoated UV grade fused silica window at 632 nm.

5.6   The internal transmittance is given by

$$T_I = e^{-\alpha t_w} \tag{A5.6a}$$

Substituting values of 3.57 cm^{-1} for α and 0.2 cm for t_w in Equation A5.6a, we obtain a value of 49% for the internal transmittance of the Ge window at 12 μm. The external transmittance is given by

$$\frac{(1-R)^2}{1-R^2 e^{-2\alpha t_w}} e^{-2\alpha t_w} \tag{A5.6b}$$

Substituting values of 0.36 for R, 3.57 cm^{-1} for α, and 0.2 cm for t_w in Equation A5.6b, we obtain a value of 20.7% for the external transmittance of the Ge window at 12 μm.

5.7   The internal transmittance is given by

$$T_I = e^{-\alpha t_w} \tag{A5.7a}$$

Substituting values of 5.28 cm^{-1} for α and 0.2 cm for t_w in Equation A5.7a, we obtain a value of 35% for the internal transmittance of the CaF$_2$ window at 9 μm.

The single-surface reflectance is given by

$$R = \left(\frac{n-1}{n+1}\right)^2 \tag{A5.7b}$$

Substituting a value of 1.327 for n in Equation A5.7b, we obtain a value of 2% for R. The external transmittance is given by

$$\frac{(1-R)^2}{1-R^2 e^{-2\alpha t_w}} e^{-2\alpha t_w} \tag{A5.7c}$$

Using values of 0.02 for R, 5.28 cm^{-1} for α, and 0.2 cm for t_w in Equation A5.7c, we obtain a value of 33.6% for the external transmittance of the CaF$_2$ window at 9 μm.

5.8   The reflectance of a window coated with a quarter-wave thick film of refractive index n_f is given by

$$R = \frac{(n_w - n_f^2)^2}{(n_w + n_f^2)^2} \tag{A5.8a}$$

where n_w is the refractive index of the window material. Note that $R = 0$, which is an ideal AR-coated window if

$$n_f = \sqrt{n_w} \qquad (A5.8b)$$

The value of n_w is 4.0 for a Ge window in the 3–12 μm range. Therefore, the value of n_f for ideal AR coating is 2.0. The refractive index of TiO_2 is 2.55, which is close to the value of 2.0. The reflectance of a germanium window AR-coated with a quarter-wave thick film of TiO_2 will be 5.7%, which is significantly lower than the value of 36% for the uncoated Ge window.

5.9   The value of the refractive index of CaF_2 windows at 532 nm is 1.435. MgF_2 with refractive index of 1.35 is commonly used for AR coating of glass windows. The reflectance of a CaF_2 window AR-coated with a quarter-wave thick film of MgF_2 will be 1.4%.

5.10   The value of the refractive index of an Al_2O_3 window at 532 nm is 1.772. The square root of 1.772 is 1.33, which is very close to the refractive index of MgF_2. The reflectance of an Al_2O_3 window AR-coated with a quarter-wave thick film of MgF_2 will be 0.02%, which is really close to zero.

5.11   The parallel displacement of the light beam in passing through a window is given by

$$\delta = \frac{\sin(\theta_i - \theta_t)}{\cos \theta_t} t_w = \sin \theta_i \left(1 - \frac{\sqrt{1 - \sin^2 \theta_i}}{\sqrt{n^2 - \sin^2 \theta_i}} \right) t_w \qquad (A5.11)$$

Substituting values of 30° for θ_i 1.5 for n, and 10 mm for t_w in Equation A5.11, we obtain a value of 1.94 mm for δ.

6.1   The external transmittance is given by

$$T_E = \frac{(1 - R)^2}{1 - R^2 T_i^2} T_i \qquad (A6.1a)$$

$$R = \left(\frac{n - 1}{n + 1} \right)^2 \qquad (A6.1b)$$

Using a value of 1.5 for n in Equation A6.1b, we obtain a value of 4% for R. Optical density is given by

$$OD = \log \left(\frac{1}{T_E} \right) \qquad (A6.1c)$$

Calculated values of T_E and OD are shown in the following Table A6.1.

TABLE A6.1
Values of T_E and O_D

λ (nm)	350	400	450	500	550	600
T_i	0.352	0.809	0.896	0.932	0.821	0.390
T_E	3.24×10^{-1}	7.46×10^{-1}	8.27×10^{-1}	8.60×10^{-1}	7.57×10^{-1}	3.60×10^{-1}
OD	0.49	0.13	0.08	0.07	0.12	0.44
λ (nm)	650	700	750	800	850	900
T_i	6.4×10^{-2}	4.3×10^{-3}	2.8×10^{-4}	4.6×10^{-5}	2.5×10^{-5}	3.6×10^{-5}
T_E	5.90×10^{-2}	3.96×10^{-3}	2.58×10^{-4}	4.24×10^{-5}	2.30×10^{-5}	3.32×10^{-5}
OD	1.23	2.40	3.59	4.37	4.64	4.48

Plots of T_E and OD versus λ are shown in Figure A6.1 and Figure A6.2, respectively.

6.2 *OD* is proportional to the thickness of the filter. The value of OD for a 1.0-mm thick Schott BG60 glass filter at 600 nm is 0.44. Therefore, the value of OD for a 0.1-mm thick Schott BG60 glass filter at 600 nm is 0.044.

6.3 External transmittance is given by

$$T_E = \frac{(1-R)^2}{1-R^2 T_i^2} T_i \tag{A6.3}$$

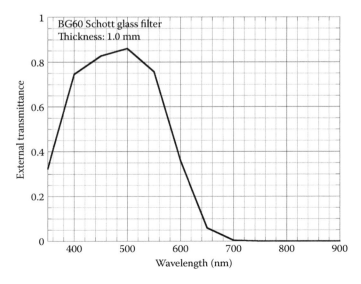

FIGURE A6.1 External transmittance of BG60 Schott glass filter.

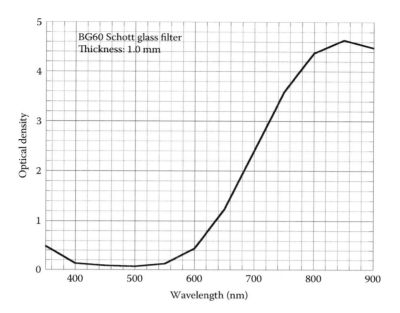

FIGURE A6.2 Optical density of BG60 Schott glass filter.

TABLE A6.2
External Transmittance T_E

λ (nm)	510	520	530	540	550	560
T_E	4.78×10^{-3}	0.125	0.473	0.738	0.838	0.885
λ (nm)	570	580	590	600	610	
T_E	0.899	0.905	0.907	0.908	0.908	

where the value of R is 0.041. Calculated values of the external transmittance are shown in Table A6.2.

A plot of T_E versus λ is shown in Figure A6.3.

6.4 Calculated values of external transmission of a 1.0-mm thick Schott UG1 bandpass filter are shown in Table A6.3.

A plot of T_E versus λ is shown in Figure A6.4.

6.5 Optical density is given by

$$OD = \log\left(\frac{1}{T_E}\right)$$ (A6.5)

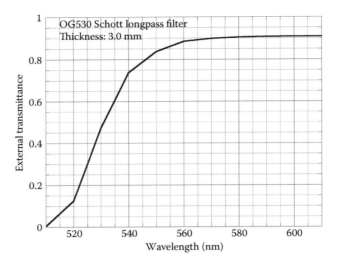

FIGURE A6.3 External transmittance of OG530 Schott longpass filter.

TABLE A6.3

External Transmittance T_E of a 1.0-mm Thick Schott UG1 Bandpass Filter

λ (nm)	290	300	310	320	330	340	350
T_E	0.0334	0.14	0.303	0.470	0.596	0.680	0.731
λ (nm)	360	370	380	390	400	410	420
T_E	0.754	0.735	0.640	0.396	0.125	0.0154	0.000777

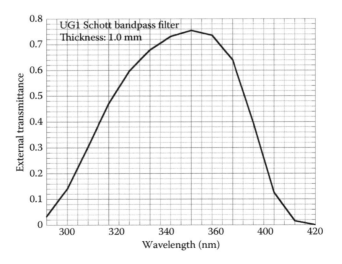

FIGURE A6.4 External transmittance of UG1 Schott bandpass filter.

TABLE A6.4

Values of OD for 500 nm Thorlabs Dielectric Bandpass Filter

λ (nm)	490	491	492	493	494	495	496
OD	2.52	2.15	1.85	1.49	1.12	0.801	0.558
λ (nm)	497	498	499	500	501	502	503
OD	0.407	0.336	0.315	0.312	0.313	0.332	0.413
λ (nm)	504	505	506	507	508	509	510
OD	0.600	0.951	1.29	1.68	2.05	2.30	2.70

Calculated values of OD are shown in Table A6.4 for Thorlabs dielectric bandpass filter with CWL of 500 ± 2 nm and FWHM of 10 ± 2 nm.

A plot of OD versus λ is shown in Figure A6.5.

6.6 Suppose CWL of the bandpass filter is λ_B and FWHM is $\Delta\lambda_B$. Wavelengths of the longpass and shortpass filters are given by

$$\lambda_L = \lambda_B - \frac{1}{2}\Delta\lambda_B$$
$$\lambda_S = \lambda_B + \frac{1}{2}\Delta\lambda_B$$

(A6.6)

6.7 Optical density is given by

$$OD = \log\left(\frac{1}{T_E}\right)$$

(A6.7)

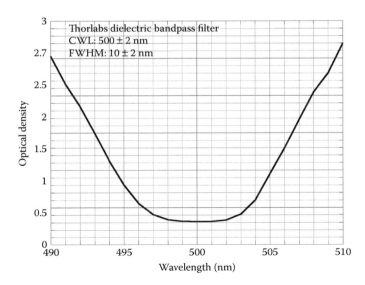

FIGURE A6.5 OD of 500-nm bandpass filter.

Calculated values of OD are given in Table A6.5 for Thorlabs 594 ± 2 nm notch filter.

A plot of OD versus λ is shown in Figure A6.6.

6.8   Optical density is given by

$$OD = \log\left(\frac{1}{T_E}\right) \tag{A6.8}$$

TABLE A6.5

Values of OD for the 594±2 nm Notch Filter

λ (nm)	581	582	583	584	585	586	587
OD	0.0167	0.0302	0.155	0.599	1.89	5.00	5.00
λ (nm)	588	589	590	591	592	593	594
OD	5.00	5.00	5.00	5.00	5.00	5.00	5.00
λ (nm)	595	596	597	598	599	600	601
OD	5.00	5.00	5.00	5.00	5.00	3.96	0.860
λ (nm)	602	603	604	605	606	607	608
OD	0.477	0.286	0.0294	0.0228	0.0138	0.0115	0.00937

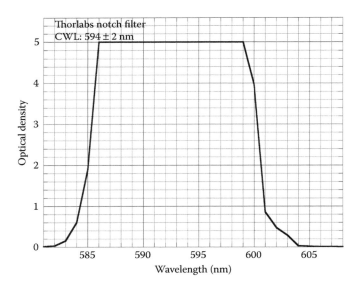

FIGURE A6.6   Optical density of 594-nm notch filter.

TABLE A6.6
Values of OD for the Semrock 532 nm Longpass Raman Filter

λ (nm)	532.0	532.2	532.4	532.6	532.8	533.0	533.2
OD	6.00	5.26	4.31	4.24	4.17	4.02	3.72
λ (nm)	533.4	533.6	533.8	534.0	534.2	534.4	534.6
OD	3.48	3.21	2.40	1.07	0.269	0.153	0.0721
λ (nm)	534.8	535.0	535.2	535.4	535.6	535.8	536.0
OD	0.0615	0.0400	0.0237	0.0232	0.0205	0.0141	0.0137
λ (nm)	536.2	536.4	536.6	536.8	537.0	537.2	537.4
OD	0.0150	0.0132	0.0101	0.00877	0.0106	0.0123	0.0123

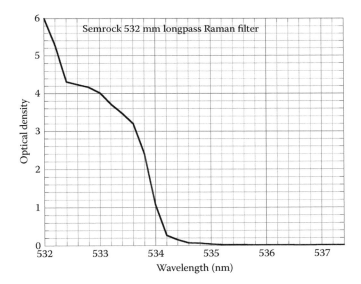

FIGURE A6.7 Optical density of 532 nm longpass Raman filter.

Calculated values of OD for the Semrock 532 nm longpass Raman filter are shown in Table A6.6.

A plot of OD versus λ is shown in Figure A6.7.

6.9 Optical density is given by

$$OD = \log\left(\frac{1}{T_E}\right) \tag{A6.9}$$

Calculated values of OD for the Semrock 532 nm shortpass Raman filter are given in Table A6.7.

A plot of OD versus λ is shown in Figure A6.8.

TABLE A6.7

Values of OD for Semrock 532 nm Shortpass Raman Filter

λ (nm)	525.0	525.2	525.4	525.6	525.8	526.0	526.2
OD	0.0141	0.0146	0.0141	0.0137	0.0137	0.0141	0.0150
λ (nm)	526.4	526.6	526.8	527.0	527.2	527.4	527.6
OD	0.0164	0.0177	0.0195	0.0218	0.0255	0.0311	0.0560
λ (nm)	527.8	528.0	528.2	528.4	528.6	528.8	529.0
OD	0.199	0.587	1.21	1.85	2.40	2.80	3.08
λ (nm)	529.2	529.4	529.6	529.8	530.0	530.2	530.4
OD	3.23	3.39	4.17	5.00	5.44	5.72	6.07
λ (nm)	530.6	530.8	531.0	531.2	531.4	531.6	531.8
OD	6.32	6.43	6.52	6.64	6.80	7.02	7.54

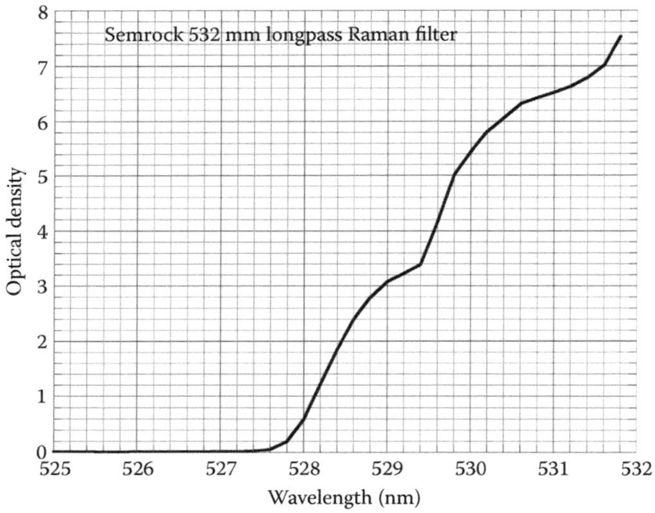

FIGURE A6.8 OD of 532 nm shortpass Raman filter.

7.1 The displacement of the transmitted beam in passing through the Ge beamsplitter is given by

$$\delta = \frac{\sin(\theta_i - \theta_t)}{\cos\theta_t} t_B = \sin\theta_i \left(1 - \frac{\sqrt{1 - \sin^2\theta_i}}{\sqrt{n^2 - \sin^2\theta_i}} \right) t_B \qquad (A7.1)$$

Substituting values of 45° for θ_i, 4.0 for n of Ge, and 1.0 mm for t_B in Equation A7.1 we obtain a value of 0.58 mm for δ.

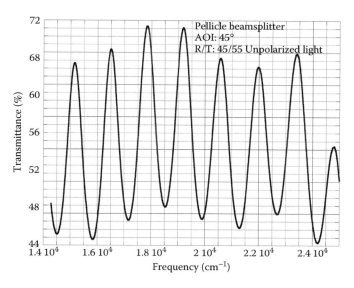

FIGURE A7.1 Transmittance of pellicle 45/55 beamsplitter.

7.2 a. Substituting values of 30° for θ_i and 1.5 for n in Equation A7.1, we obtain a value of $0.19t_B$ for δ.

b. Substituting values of 40° for θ_i and 1.5 for n in Equation A7.1, we obtain a value of $0.28t_B$ for δ.

c. Substituting values of 50° for θ_i and 1.5 for n in Equation A7.1, we obtain a value of $0.38t_B$ for δ.

d. Substituting values of 60° for θ_i and 1.5 for n in Equation A7.1, we obtain a value of $0.51t_B$ for δ.

7.3 The transmittance spectrum of a Thorlabs pellicle beamsplitter with R/T of 45/55 for unpolarized light of a 45° AOI is shown in Figure A7.1.

Fringe spacing is equal to $1.37 \times 10^3 \, \text{cm}^{-1}$, which corresponds to an optical thickness of 3.6 μm.

8.1 Let us define a dimensionless parameter

$$x = \frac{hc}{\lambda k_B T} \tag{A8.1a}$$

Equation A8.1a for Planck's radiation law may be written in terms of x as:

$$I(\lambda, T) = \frac{2\pi (k_B T)^5}{h^4 c^3} \left(\frac{x^5}{e^x - 1} \right) \Delta\lambda \tag{A8.1b}$$

Differentiating Equation A8.1b with respect to x and setting the result equal to zero, we find that $I\,(\lambda,\,T)$ is maximum for x equal to 4.965. This implies that the wavelength at which I $(\lambda,\,T)$ is maximum is given by

$$\lambda_{max} = \frac{hc}{4.965k_BT} \qquad\qquad (A8.1c)$$

Equation A8.1c yields the desired result

$$\lambda_{max}T = 2898 \ \mu m.\,K \qquad\qquad (A8.1d)$$

8.2 The wavelength for peak intensity of a blackbody source at temperature T is given by

$$\lambda_p\left(\mu m\right) = \frac{2898}{T\left(K\right)} \qquad\qquad (A8.2a)$$

Substituting a value of 1500 K for T in Equation A8.2a, we obtain a value of 1.932 μm for λ_p. The intensity/nm at 1.932 μm is given by

$$I\left(1.932\,\mu m, 1500\,K\right) = \frac{2\pi hc^2}{\left(1.932\times 10^{-4}\mathrm{cm}\right)^5}\left(\frac{1}{e^{4.965}-1}\right)\left(1.0\times 10^{-7}\,\mathrm{cm}\right)$$

Substituting values of 6.626 × 10⁻³⁴ J.s for h and 3 × 10¹⁰ cm/s for c in Equation A8.2b, we obtain a value of 9.8 mW/cm² for the intensity/nm at 1.932 μm at 1500 K.

8.3 The efficiency of the 10 W LED is 6x higher than that of the 60 W tungsten bulb. The efficiency of the tungsten bulb in the visible is 8%. Therefore, the efficiency of the 10 W LED is 48%.

8.4 The cost of electricity for operation of a 60-W incandescent bulb for 10 years is equal to 60 W × 10 h/day × 365 days/year × 10 years x $0.20/kWh, which is equal to $438.00. 11 bulbs are needed for operation for 10 years. The cost of the replacement bulbs is 11 bulbs × $1.00/bulb, which is equal to $11.00. Hence the total cost of operation of the 60-W incandescent bulb for 10 years is equal to $449.00.

8.5 The cost of electricity for operation of a 9-W LED bulb for 10 years is equal to 8 W × 10 h/day × 365 days/year × 10 years × $0.20/kWh, which is equal to $65.70. The cost of the single LED bulb with lifetime > 10 years is $10.00. Hence the total cost of operation of the 9-W LED bulb for 10 years is equal to $75.70, which is 17% of that of a 60-W incandescent bulb.

8.6 The average separation between the CO_2 laser lines in the 9.6 μm band are determined to be 2.04 cm⁻¹ by dividing the difference between the frequencies 1063.2 and 1012.3 cm⁻¹ of the outer modes by 25, which is the number of lines minus 1 in the CO_2 laser spectrum in Figure 8.5.

The average separation between the CO_2 laser lines in the 10.6 μm band are determined to be 1.88 cm^{-1} by dividing the difference between the frequencies 959.4 and 914.4 cm^{-1} of the outer modes by 24, which is the number of lines minus 1 in the CO_2 laser spectrum in Figure 8.5.

8.7   The laser output power is given by

$$P_L = \frac{\ln\left(R_1 R_2\right)^{-1/2}}{\alpha L + \ln\left(R_1 R_2\right)^{-1/2}} \left(\frac{\tau_{nr}}{\tau_{sp} + \tau_{nr}}\right)\left(P_P - P_{th}\right) \tag{A8.7}$$

Substituting values of 1.00 for R_1, 0.95 for R_2, 0.02 for αL, 0.80 for $\tau_{nr}/(\tau_{sp}+\tau_{nr})$, 10.0 W for P_P, and 1.0 W for P_{th} in Equation A8.7, we obtain a value of 4.0 W for P_L.

9.1   NEP of a detector is given by

$$NEP = \frac{\sqrt{A_d \Delta v}}{D *} \tag{A9.1}$$

where:
   A_d is the area of the detector
   Δv is the frequency bandwidth

Substituting values of $1 \times 10^{-2}\,cm^2$ for A_d, 0.5 Hz for Δv corresponding to integration time of 1.0 s, and $8 \times 10^{11}\,cmHz^{1/2}/W$ for $D*$ in Equation A9.1, we obtain a value of $8.8 \times 10^{-14}\,W$ for NEP.

9.2   The light signal is given by

$$S_L = NEP * SNR \tag{A9.2a}$$

where SNR is the signal-to-noise ratio. Substituting values of $8.8 \times 10^{-14}\,W$ for NEP and 10 for SNR in Equation A9.2a, we obtain a value of $8.8 \times 10^{-14}\,W$ for S_L. A 10-μm photon has energy

$$E_L = hv = \frac{hc}{\lambda} \tag{A9.2b}$$

where λ is the wavelength of light. Substituting values of $6.63 \times 10^{-34}\,J.s$ for h, $3 \times 10^8\,m/s$ for c, and 10 μm for λ in Equation A9.2b, we obtain a value of $1.99 \times 10^{-20}\,J$ for E_L. The number of 10-μm light photons/s incident upon the detector is given by

$$N_L = \frac{S_L}{E_L} \tag{A9.2c}$$

Substituting values of $8.8 \times 10^{-14}\,W$ for S_L and $1.99 \times 10^{-20}\,J$ for E_L in Equation A9.2c, we obtain a value of 4.4×10^7 photons/s for N_L.

9.3   NEP of the thermocouple for an integration time of $10s$ is given by

$$\mathrm{NEP}_{10s} = \frac{\mathrm{NEP}}{\sqrt{\tau}} \qquad\qquad \text{(A9.3a)}$$

Substituting values of 3×10^{-10} W/Hz$^{1/2}$ for NEP and $10s$ for τ in Equation A9.3a, we obtain a value of 0.95×10^{-10} W for NEP_{10s}. Signal power is given by

$$S_P = \mathrm{NEP}_{10s} * \mathrm{SNR} \qquad\qquad \text{(A9.3b)}$$

Substituting values of 0.95×10^{-10} W for NEP_{10s} and 3 for SNR in Equation A9.3b, we obtain a value of 2.85×10^{-10} W for S_P.

9.4   NEP of the pyroelectric detector for an integration time of $30s$ is given by

$$\mathrm{NEP}_{30s} = \frac{\mathrm{NEP}}{\sqrt{\tau}} \qquad\qquad \text{(A9.4a)}$$

Substituting values of 1×10^{-8} W/Hz$^{1/2}$ for NEP and $30s$ for τ in Equation A9.4a, we obtain a value of 1.83×10^{-9} W for NEP_{30s}. Signal power is given by

$$S_P = \mathrm{NEP}_{30s} * \mathrm{SNR} \qquad\qquad \text{(A9.4b)}$$

Substituting values of 1.83×10^{-9} W for NEP_{30s} and 10 for SNR in Equation A9.4b, we obtain a value of 1.83×10^{-8} W for S_P.

9.5   NEP is given by

$$\mathrm{NEP} = \frac{\sqrt{A_d/2\tau}}{D*} \qquad\qquad \text{(A9.5)}$$

Substituting values of 4.0×10^{-3} cm^2 for A_d, 10 s for τ, and 4.0×10^{12} cm Hz$^{1/2}$/w for $D*$ in Equation A9.5, we obtain a value of 3.5×10^{-15} W for NEP.

9.6   Photocurrent is given by

$$i = R_\lambda P_\lambda \qquad\qquad \text{(A9.6)}$$

Substituting values of 0.7 A/W for R_λ and 1×10^{-13} W for P_λ in Equation A9.6, we obtain a value of 0.07 pA for i.

9.7   The ionization coefficient for the electrons is given by

$$\alpha_n = \frac{1-(1/M)}{w} \qquad\qquad \text{(A9.7)}$$

Substituting a value of 100 for M in Equation A9.7, we obtain a value of $0.99/w$ for α_n.

9.8   The multiplication factor for the PMT is given by

$$M = \delta^N \tag{A9.8}$$

Substituting values of 5 for δ and 10 for N in Equation A9.8, we obtain a value of 9.8×10^6 for M.

9.9   The signal current is given by

$$i_s = \eta e \frac{P_s}{h\nu} \tag{A9.9}$$

where:
    η is the quantum efficiency
    e is the electronic charge
    P_s is the signal power
    $h\nu$ is the photon energy

Substituting values of 0.70 for η, 1.6×10^{-19} C for e, 1 pW for P_s, and 3.315×10^{-19} J for $h\nu$ for 600 nm photons in Equation A9.9, we obtain a value of 0.34 pA for i_s.

9.10   The shot noise of the photodetector is given by

$$i_{ns} = e\sqrt{\eta P_S / \tau} \tag{A9.10}$$

Substituting values of 1.6×10^{-19} C for e, 0.70 for η, 1 pW for P_S, and 10 s for τ in Equation A9.10, we obtain a value of 4.2×10^{-26} A for i_{ns}.

References

Bell, W. E., *App. Phys. Lett.*, **4**, 34 (1964).

Bennett, W. R., Jr., W. L. Faust, R. A. McFarlane, and C. K. N. Patel, *Phys. Rev. Lett.,* **8**, 470 (1962).

Boyle, W. S. and K. F. Rodgers, Jr., *J. Opt. Soc. Am.*, **49**, 66 (1959).

Bridges, W. B. and A. N. Chester, *Appl. Optics*, **4**, 573 (1965).

Bridges, W. B., *Proc. IEEE*, **52**, 843 (1964).

Bridges, W., *Appl. Phys. Lett.*, **4**, 128 (1964).

Chen, C. G. and M. L. Schattenburg, A brief history of gratings and the making of the MIT nanoruler, *MIT News*, January 28, 2004.

Cooper, J., *Rev. Sci. Instrum.*, **33**, 92 (1962).

Corsi, S., G. Dall'Oglio, and F. Melchiorri, *Infrared Phys.*, **13**, 253 (1973).

Edmund Optics, www.edmundoptics.com/optics, 2017.

First Sensor, Introduction to silicon photomultipliers (SiPMs), Version 03-12-15, www.first-sensor.com, 2015.

Fowles, G. R. and R. C. Jensen, *Appl. Optics*, **3**, 1191 (1964).

Fowles, G. R., *Introduction to Modern Optics*, Holt, Rinehart and Winston, New York (1975).

Fowles, G. R., W. T. Silfvast, and R. C. Jensen, *IEEE J. Quant. Electron.*, **QE-1**, 183 (1965).

Gandy, H. W. and R. J. Ginther, *Appl. Phys. Lett.*, **1**, 25 (1962).

Geusic, J. E., H. M. Marcos, and L. G. Van Uitert, *Appl. Phys. Lett.*, **4**, 182 (1964).

Hamamatsu Photonics, Handbook, Chapter 2, Basic principles of photomultiplier tubes, 2007.

Harris, D. C., *Materials for Infrared Windows and Domes: Properties and Performance*, SPIE Optical Engineering Press, Bellingham, Washington DC (1999).

Harrison, G. R. and G. W. Stroke, *J. Opt. Soc. Am.*, **45**, 112 (1955).

Harrison, G. R., *J. Opt. Soc. Am.*, **39**, 413 (1949).

Hecht, E. and A. Zajac, *Optics*, Addison-Wesley Publishing Company, Boston, MA (1974).

Holonyak, N., Jr. and S. F. Bevacqua, *Appl. Phys. Lett.*, **1**, 82 (1962).

ISP Optics, www.ispoptics.com, 2017.

Javan, A., W. R. Bennett, and D. R. Harriot, *Phys. Rev. Lett.*, **6**, 106 (1961).

Jenkins, F. A. and H. E. White, *Fundamentals of Optics*, 4th ed., McGraw-Hill Book Company, New York (1976).

Johnson, L. F. and R. A. Thomas, *Phys. Rev.*, **131**, 2038 (1963).

Johnson, L. F., G. D. Boyd, and K. Nassau, *Proc. IRE.*, **50**, 87 (1962a).

Johnson, L. F., G. D. Boyd, K. Nassau, and R. R. Soden, *Phys. Rev.*, **126**, 1406 (1962b).

Keefe, W. M. and W. J. Graham, *Appl. Phys. Lett.*, **7**, 263 (1965).

Kingston, R. H., *Optical Sources, Detectors, and Systems*, Academic Press, Cambridge, MA (1995).

Kiss, Z. J. and R. C. Duncan, Jr., *Proc. IRE*, **50**, 1531 (1962).

Krupke, W. F. and J. B. Gruber, *J. Chem. Phys.*, **41**, 1225 (1964).

Kruse, P. W., L. D. McGlauchlin, and R. B. McQuistan, *Elements of Infrared Technology*, John Wiley & Sons, New York (1962).

Land, E. H., *J. Opt. Soc. Am.*, **41**, 957 (1951).

Lerner, J. M., J. Flamand, J. P. Laude, G. Passereau, and A. Thevenon, *SPIE*, **240**, 82 (1980).

Liston, M. O., *J. Opt. Soc. Am.*, **37**, 515A (1947).

Low, F. J., *J. Opt. Soc. Am.*, **51**, 1300 (1961).

Ludlow, J. H., W. H. Mitchell, E. H. Putley, and N. Shaw, *J. Sci. Instrum.*, **44**, 694 (1967).

Maiman, T. H., *Nature*, **187**, 493 (1960).

Mathias, L. E. S. and J. T. Parker, *Appl. Phys. Lett.*, **3**, 16 (1963).

McFarlane, R. A., *Appl. Optics*, **3**, 1196 (1964b).

McFarlane, R. A., *Appl. Phys. Lett.*, **5**, 91 (1964a).

Melles Griot, www.mellesgriot.com, 2017.

Motes, R. A. and R. W. Berdine, *Introduction to High-Power Fiber Lasers,* Directed Energy Professional Society, New York (2009).

Moulton, P. F., *Opt. Soc. Am.*, **B3**, 125 (1986).

Newport Corporation, www.newport.com, 2017.

Nikon, www.nikonins.org, 2017.

NIST, www.nist.gov/PhysRefData, 2017.

O'Brien, P., MIT Lincoln Laboratory, Lexington, MA, 02420, 2016.

Paananen, R. A., C. L. Tang, and F. A. Horrigan, *Appl. Phys. Lett.*, **3**, 154 (1963).

Patel, C. K. N. and R. J. Kerl, *Appl. Phys. Lett.*, **5**, 81 (1964).

Patel, C. K. N., *Phys. Rev.*, **136**, 1187 (1964).

Patel, C. K. N., R. A. McFarlane, and W. L. Faust, *Phys. Rev.*, **133**, A1244 (1964).

Patel, C. K. N., W. R. Bennett, Jr., W. L. Faust, and R. A. McFarlane, *Phys. Rev. Lett.*, **9**, 102 (1962).

Payne, S. A., J. A. Caird, L. L. Chase, L. K. Smith, N. D. Nielsen, and W. F. Krupke, *J. Opt. Soc. Am.*, **B8**, 726 (1991).

Perkin-Elmer, www.perkin-elmer.com, 2017.

Perry, M. D., R. D. Boyd, J. A. Britten, D. Decker, B. W. Shore, C. Shannon, and E. Shults, *Optics Lett.*, **20**, 940 (1995).

Pollack, S. A., *Proc. IEEE*, **51**, 1793 (1963).

Refractive Index, www.refractiveindex.info, 2017.

Rigden, J. D. and A. D. White, *Nature*, **198**, 774 (1963).

Rosenbluh, M., MS thesis, Design and evaluation of a continuous-wave, step-tunable far infrared source for solid state spectroscopy, MIT, Cambridge, MA (1975).

Sabisky, E. S. and H. R. Lewis, *Proc. IEEE*, **51**, 53 (1963).

Schott, www.us.schott.com, 2017.

Semrock, www.semrock.com, 2017.

Solomon, R. and L. Mueller, *Appl. Phys. Lett.,* **3**, 135 (1963).

Sorokin, P. P. and M. J. Stevenson, *IBM J. Res. Develop.*, **5**, 56 (1961).

Sorokin, P. P. and M. J. Stevenson, *Phys. Rev. Lett.,* **5**, 557 (1960).

Sorokin, P. P., M. J. Stevenson, J. R. Lankard, and G. D. Pettit, *Phys. Rev.*, **127**, 503 (1962).

Spencer, D. J., T. A. Jacobs, H. Mirels, and R. W. F. Gross, *Int. J. Chem. Kinetics*, **1**, 493 (1969).

Sustainable Supply, www.sustainablesupply.com, 2016.

Tatian, B., *Appl. Optics*, **23**, 4477 (1984).

Thorlabs, www.thorlabs.com, 2017.

VisionTech Systems, www.visiontechsystems.com, 2017.

Waard, R. D. and E. M. Wormser, *Proc. IRE*, **47**, 1508 (1959).

Wallace, R. J., *Astrophy. J.*, **22**, 123 (1905).

Waynant, R. W. and P. H. Klein, *Appl. Phys. Lett.*, **46**, 14 (1985).

Wickersheim, K. A. and R. A. Lefever, *J. Electrochem. Soc.*, **111**, 47 (1964).

Wikipedia, www.wikipedia.org, 2017.

Wittke, J. P., Z. J. Kiss, R. C. Duncan, Jr., and J. J. McCormick, *Proc. IEEE*, **51**, 56 (1963).

Yariv, A., S. P. S. Porto, and K. Nassau, *J. Appl. Phys.*, **33**, 2519 (1962).

Index

Note: Tables are indicated by **bold** page references; figures, photos and/or illustrations are indicated by an *italicized* number.